The Global City and the Holy City

The Global City and the Holy City

Narratives on Knowledge, Planning and Diversity

Tovi Fenster

PEARSON
Prentice
Hall

Harlow, England • London • New York • Boston • San Francisco • Toronto
Sydney • Tokyo • Singapore • Hong Kong • Seoul • Taipei • New Delhi
Cape Town • Madrid • Mexico City • Amsterdam • Munich • Paris • Milan

Pearson Education Limited
Edinburgh Gate
Harlow
Essex CM20 2JE
England

and Associated Companies throughout the world

Visit us on the World Wide Web at:
www.pearsoned.co.uk

First published 2004

ISBN 0 582 35660 1

British Library Cataloguing-in-Publication Data
A catalogue record for this book is available from the British Library

Library of Congress Cataloging-in-Publication Data
Fenster, Tovi.
    The global city and the holy city : narratives on knowledge, planning, and diversity / Tovi
Fenster.
        p. cm.
    Includes bibliographical references and index.
    ISBN 0-582-35660-1 (pbk.)
    1. Quality of life—England—London.   2. Human comfort—England—London.
3. Quality of life—Jerusalem.   4. Human comfort—Jerusalem.   I. Title.

HN398.L7F46 2004
306'.09421'2—dc21
                                                                                2003053587

Printed and bound in Great Britain by
CPI Antony Rowe, Chippenham and Eastbourne

Transferred to Digital Print on Demand 2010

Typeset in 10/12 pt Sabon by 35
Produced by Pearson Education Malaysia Sdn Bhd,

Contents

List of figures

All photos were taken by Tovi Fenster unless otherwise credited:

List of tables

Preface

The inspiration to write this book came out of my personal experiences of living in my two home cities: London and Jerusalem. Jerusalem is my home 'address', reflecting many aspects and expressions of my identity. London is my emotional and intellectual home. It is where during my years of working on my PhD studies I have learned that a city can become home not so much in official terms but in everyday practices. My life between the two cities continues today, in Jerusalem as a permanent home and in London through visits several times a year.

This book is based in many ways on these two home networks. I wish to dedicate this part of it to those colleagues and friends who were willing to share time and space to reflect on the different ideas that came out in the process of my writing it. In Jerusalem and Israel I wish to thank Haim Yacobi for the hours of discussions on the various topics of the book and for his helpful and constructive comments on some of the early draft chapters. I wish to thank Nurit Alfasi, who has been so supportive and patient and has made very helpful comments. My thanks go to Yitzhak Omer for clarifications and helpful comments on Part III, and to Oren Yiftachel for his support and comments. Special thanks to Shamay Assif, whose ideas of comfort, belonging and commitment inspired me to take them further. I thank him for the time he spent discussing these issues and for his willingness to pursue a dialogue with me on the various topics that emerged in the process of my writing.

In London, I wish to thank Steve Pile for his support and helpful comments on some of the draft chapters. Thanks too to Nadia Taher who helped me to sharpen some of the arguments in some of the chapters. I also wish to thank Micheal Safier, Derek Diamond, Caren Levy and Micheal Parks for sharing their ideas with me and helping me to understand the planning system in London; Mr Nisar Ahmed, who dedicated time to clarify the intricacies of the life of the Bangladeshi in London; and Kay Jordan, Roda Brawne and Lea Gratch of the Spitalfields Small Business Association for presenting a different perspective and providing helpful insights into events in the Spitalfield – Banglatown area.

I am grateful to my friends and colleagues in other parts of the world who spent time reading and commenting on early drafts of the book, especially to

Leoni Sandercock for her kind and useful comments, to Maria Dolors Garcia Ramon for her supportive and constructive remarks and to Janice Monk for her insightful ideas.

I wish to thank all the people in London and Jerusalem who were willing to share with me their local knowledge of their daily life and who enriched me so much with their experiences, attitudes, emotions and wisdom. I dedicate this book to them.

Last but not least, thanks go to my family members, who have always been a stable source of support and love, especially to our young generation: Chen, Tom, Shir, Noam, Nir and Neta, who taught me the essence of comfort, belonging and commitment.

CHAPTER 1

Introduction

What does it mean to live in a city today? Do daily practices in cities have their similarities in spite of the different histories, economies, politics and cultures that cities represent? This is the first question that I pose in this book, perhaps because of my own personal experiences of living everyday life in various cities in the world and being able to feel 'at home' in some of them. These experiences made me interested in what makes a city feel like a home and whether this is common to many people in many cities. So I pose another question: are the assumed differences between London, the global city and Jerusalem, the holy city reflected in people's experiences of living in the two cities? I have chosen to focus on the daily practices of people in two specific cities in order to explore the similarities and differences of life in these cities. The book suggests that some of these everyday practices are not so different as might be assumed. It proposes that people of different national, cultural or gender identities might experience their city-as-home for similar reasons.

I then wanted to examine the links that can be established between these everyday experiences, what I term 'local embodied knowledge', and planning practice. This I perceive as the end result and as one of the major challenges of the book: how to articulate the local knowledge that has been exposed and analysed in the various chapters into planning practice. The book suggests new ways of incorporating these similar and different experiences in the planning process. The 'planning process' is perceived here as a joint procedure integrating both the professional knowledge of planners and local embodied knowledge of people in shaping and reshaping cityscapes.

Why London and Jerusalem? As I already mentioned in the Preface, I have chosen London and Jerusalem as the focus of this research because the two cities are my home. It is in Jerusalem where I spend my everyday life as my resident home – city. Jerusalem is my home by choice. I was born in Tel Aviv and lived there most of my childhood and early adult life. My current work is in Tel Aviv but my choice for home is Jerusalem. Why? This is perhaps one of the deep and emotional reasons for choosing this topic for research. I want to find out what makes Jerusalem a home for people of diverse identities, especially in the light of its image as one of the holiest cities in the world, a place of symbolism for Muslims, Christians and Jews. And also a city that is associated with rigidity, perhaps fanatics, strict rules and boundaries between what is permitted and what is forbidden in daily life.

London is where I grew intellectually during my PhD studies and in the almost ritualistic and repetitive visits I make several times each year. It is where I established my own community and networks and where I have learnt that 'home' is not only an address but also an embodied and deep sentiment of attachment and belonging to a place. I wanted to find out how a non-English person such as myself can feel so much at home in a city famous for its impact on globalisation and its images of cosmopolitanism, openness and tolerance but also for its negative and depressing connotations, especially for non-English and other types of 'aliens'.

It is perhaps because of these contrasting images and personal perceptions of the two cities that I have chosen to focus my research on them. It is precisely to learn from other people's everyday practices what makes a city a home and how we can use this knowledge in order to make our cities more of a home, especially in an era of increasing influence of globalising economies and spaces. I have chosen to engage the question of a-city-as-home with three elements: a sense of comfort, a sense of belonging and a sense of commitment to the environment as people who live in the two cities perceive them.

Dealing with recent developments in these two cities two major trends can be identified: first, the effects of globalisation of capitalist economies in many parts of the world, which has been accelerated by the collapse of the socialist regimes. This has caused the spread of information technology, resulting in the increasing power of financial managers – the creation of a 'network society' (Castells 1997) in which power is in the hands of those who have the financial, technological and political resources to exploit expanding markets. Another perhaps contrasting trend that can be identified is the growing voices of the 'locale', the particular both in economies but more so as a political, social and cultural standpoint (Ram 1999). Both trends have their effects on daily practices in Jerusalem and London.

These trends are also echoed in recent alternative planning theories and methodologies. It is at this point where modernist, universalistic planning traditions begin to be challenged and questioned. Are modernist planning schemes, usually deriving from rational comprehensive planning methodology, suitable any more for planning cities that experience these sometimes contrasting trends? And what is the role of the state in these transformatory periods? States, whether rich or poor, are trying to be part of capitalist economic forces, with mainstream planning traditions as part of these trends. At the same time states are aware of the growing social, cultural, political and environmental problems in cities, of the growing socio-economic gaps between the poor and the rich, of the degradation and crime in some parts of the city and of the louder voices that demand acknowledgement in urban planning and management of the diversity of cultural, race, gender, age and sexual identities people posses.

In order to explore some of these issues the book is built on the local knowledge embodied in the everyday life experiences of people living in London and Jerusalem. Each of the topics analysed in the book is a result of my interpretation of people's knowledge, which has been created out of their identities. The book is structured and based on the same idea it aims to promote, that is, to approach local knowledge as equally valuable as professional academic knowledge and to do so both in the process of writing this book and in the planning process itself.

'Local embodied knowledge' is defined here as the ways people interpret and the meanings given to the concepts of comfort, belonging and commitment. This knowledge is 'local' as it represents the daily life of people living in various locales, from the more intimate, private spaces to the more 'public' spaces that people use in their everyday activities. It is a knowledge that is 'embodied' through their identities and bodily experiences. Incorporating these intimate experiences in the planning process is what moves the planning practice forward and makes the whole process more grounded and therefore more suitable to people's needs and, especially, desires.

Why look at these three particular elements? Because they reflect the many perspectives of people–environment relationships. Environment is taken here in its broadest sense and includes six hierarchies: home, building, street, neighbourhood, city centre and city, urban parks and transportation. In order to explore the deeper meanings of people–environment relationships I use people's interpretations and meanings of these three concepts in relation to the six hierarchies of space. Comfort, belonging and commitment can be associated with the very intimate reflections of these relationships, such as the meaning of comfort and belonging in one's own home, but they are also associated with the very 'public' aspects of these relationships, such as commitment to one's own city or country.

I first came across these terms in the National Master Plan of Israel (Ministry of the Interior 1999 *Tama/35*), where they were identified by the architect of the plan – Shamay Assif – as consisting of a 'good quality of life':

> The somehow overused term 'quality of life' means a combination of **comfort,**[1] **belonging and commitment to** the society in Israel with the ties to **the place, the identification and the commitment** with the physical environment – this is the vision of our landscape. (*Tama/35*: 2–3)

Shamay Assif explains why he chose these elements as reflecting a good quality of life:

> '*I asked myself what is a quality of life and I started to develop one of the most significant images of the home. I asked myself what is the meaning of "home" and from this . . . I . . . formulated these three concepts. Each of the concepts adds another meaning. Each of them stays by itself. Calmness or comfort is associated with loosening up, unwinding, and warmth . . . that it's yours. Belonging goes together with commitment, it's not a passive belonging but you are committed to the place you're in and this place demands it from you. Comfort and belonging are not there, you have to create them . . . in the process of making the National Master Plan these concepts were in my head, I saw [a] correlation between the home and the country.*' (29 February 2001)

It might be argued that these three elements are context related, that is, they represent a quality of life in Israel, a relatively new country still struggling to shape its national identity and formations of belonging and commitment among its citizens. But I argue that comfort, belonging and commitment have much more universal and broader meanings, which are relevant in general to people living in urban spaces especially in a period of increasing global effects and influences. I further argue that since one of the prime challenges of urban planning, development and management in the twenty-first century is to make people feel 'at home' in globalised urban spaces it is important to understand the meanings of comfort, belonging and commitment at the different levels of environment.

The analysis of these three elements is used as a means to challenge the following three questions:

1. *Are all cities alike?* Do people experience the same everyday life in such contrasting cities as the global, world city of London and the holy city of Jerusalem? Here the debate around the commonalties and differences of daily practices and local embodied knowledge of people living in cities is highlighted. Are these experiences locally embedded or are there some shared universal experiences of living in urban spaces, whether 'global' or 'holy'? In other words are the concepts 'global' and 'holy' interchangeable so that in fact both London and Jerusalem are global and holy at the same time? Analysing people's interpretations and meanings of comfort, belonging and commitment in the two cities in Part II of the book sheds some light on these debates.

2. *Do identity issues have any effect on people's everyday life experiences?* Two significant identities are at focus in the analysis; gender and national identity. The analysis of each of the concepts discovers how these identities shape people's ways of interpreting and the meanings given to comfort, belonging and commitment. The analysis is carried out among people of both 'minority' and 'majority' groups in the two cities. Nationality is an identity that cross-cuts the analysis throughout the book, as does gender identity. But the narratives themselves made me realise that gender identity deserves a special and separate analysis, which appears in Chapter 8.

3. *Is local knowledge embodied in people's daily practices valid and viable to the planning practice?* Parts III and IV challenge this debate. Part III discusses the use of cognitive maps as a methodology to deepen the understanding of the meanings of comfort, belonging and commitment and Part IV demonstrates the practicalities of incorporating local embodied knowledge into the professional knowledge of planning and provides alternative thinking that moves the discussion forward.

Methodology

In analysing and formulating the 'local embodied knowledge' I interviewed people who live in Jerusalem and London about their understandings and interpretations of the concepts comfort, belonging and commitment in order to understand their associations, feelings, perceptions, desires and needs regarding these three elements.

A qualitative-structured method of interview was used in order to establish an in-depth understanding of people's narratives. This knowledge stands as a base for the everyday life and reality of each person living in the city, which by way of generalisation becomes a body of local knowledge. This can become a 'knowledge base' for certain procedures of urban planning.

The interviews were based on a structured method of interview where questions are identified and phrased in advance. All the interviewees were asked the same questions in the same order (Patton 1987). This method decreases the bias of the interviewer and makes it easier to organise the information according to the required themes. The disadvantage of this method lies in its restricted flexibility during the interview. It reduces the possibilities for manoeuvre and ad hoc responses (Patton 1987). The questions are shown in Table 1.1.

I interviewed 66 people during the years 1999–2002. The division of interviewees by gender and city of residence appears in Table 1.2.

The selection of people to interview was conducted using the snowball technique. I started interviewing people I knew and whom I thought would be

Table 1.1 The fieldwork interview

	Your home	Your building	Your street	Your neighbourhood	Your city and city centre	Transport, parks
What is comfort in: What is a sense of belonging in: What is commitment for:						

Table 1.2 Number of interviewees by gender and city of residence

	Female	Male	Total
London	11	21	32
Jerusalem	23	11	34
Total	34	32	66

interested in taking part in such a project. Through them I got to know other people, and so on. The interviews took place wherever the interviewee preferred to meet (home, work, café, restaurant) so that they felt comfortable enough to talk. In analysing the findings I approached each answer as a narrative in itself as they reflect a very personal point of view of the interviewee, deriving from their own experiences. My way of analysis was horizontal, that is, I arranged all the narratives related to each of the questions. As already mentioned, this method allowed me to make a comparison and explore the main themes or ideas that people raised in the interviews.

In analysing the narratives (see Parts II and III) I used the narrative analysis method as elaborated in Part II. Such methodology allowed me to explore the deeper meanings of each of the concepts being focused on in the book. This qualitative method also has its own limitations, especially with regard to the relatively low number of interviewees in each city, which does not permit any generalisation and perhaps does not allow any special reference to the diversity of people living in each of the cities. The narrative analysis method can also be perceived as subjective, reflecting the researcher's own values and attitudes and her own choices of the topic to focus on and develop in the analysis. However, this method does allow the exposure of the nuances and the delicate meanings of each of the concepts that is impossible in any quantitative method, and it is for that reason I chose to use this method for my research.

In order to keep the links between people's narratives and their identities each person I interviewed was asked to label themselves according to their own nationality, ethnicity and cultural background. Their personal labelling appears in this order in every quotation I use so that their local embodied knowledge is contextualised within their own identity issues.

The people I interviewed were mostly what we label as 'upper-middle class', with the exception of some Bangladeshi people I interviewed in London and the Palestinians in Jerusalem. It might be assumed that their status affects the way they articulate the three elements.

One of the striking experiences in interviewing the Palestinians in Jerusalem and the Bangladeshis in London was that I sometimes had to change my structured system of interviewing into an in-depth, unstructured interview in order to enable the interviewees to express themselves fluently. I realised that the concepts of comfort, belonging and commitment are not always self-explanatory, nor as easy to define and discuss. This experience made me understand two basic elements that I return to in Part IV. First, the extent to which these concepts are culturally oriented, meaning that people of different cultural settings or even different socio-economic backgrounds perceive these terms differently. Second, my choice to concentrate on these specific concepts could be perceived as another expression of a 'professional–academic knowledge', which dictates and affects local embodied knowledge. At the same time, concentrating on these three elements is definitely an important part of local embodied knowledge, partial indeed, but it is part of what it means to live in the city. It so happened that, when discussing the notion of 'quality of life', most people mentioned these three concepts as part and parcel of their quality of life.

Another important insight gained from talking to people is the effect of time/space changes on the meanings of the concepts. I returned to some of my interviewees after two years and re-interviewed them. Obviously those that experienced dramatic changes in their lives, such as emigrating to another country or becoming mothers, had different interpretations and perceptions of these concepts.

Drawing cognitive maps

After the interviews, I asked people to draw mental maps of their childhood environment, their present environment and their utopian–desired environment. After they had finished their drawings I discussed with each interviewee what their views were on symbols of comfort, belonging and commitment in their own drawings (see Part III). I use cognitive mapping as another method to deepen my understanding of people's experiences and perceptions as related to their own environment. The background and methodology of mental mapping is elaborated in Part III.

The structure of the book

As already mentioned, the book aims to bring together people's narratives of their everyday practices in London and Jerusalem, narratives that are identity related. A connection is made in the book between those narratives identified as 'local embodied knowledge' and the professional planning knowledge in order to move the debates in planning practice forward.

For that purpose, the book consists of four parts, each having a specific function in analysing what the book aims to present.

Part I: Planning, Knowledge and Diversity in the City provides the conceptual framework. It highlights the key issues that the book deals with, that is, planning traditions, globalisation and the city, the discourse around knowledge and issues of identity differences and diversities and their implications in city life. It provides not only the conceptual framework but also the contextual framework of the existing planning systems in London and Jerusalem.

Part II: The Local Embodied Knowledge of Comfort, Belonging and Commitment in the Global and the Holy City deconstructs the three central elements consisting of local embodied knowledge: that of comfort (Chapter 5), belonging (Chapter 6) and commitment (Chapter 7). It provides a general analysis of the three elements by comparing people's narratives in Jerusalem and London and analysing the relationships between them in order to identify the similarities and differences in people's experiences in the two cities. In Chapter 8 the three elements are analysed with regards to gender identities.

The main goal of *Part III: Different Ways of Knowing – Diversity, Knowledge and Cognitive Temporal Maps* is to introduce yet another methodology for identifying local embodied knowledge, that of cognitive mapping. This part illustrates how by analysing cognitive maps it is possible to explore new ways of knowing and new ways of expressions, which are not only verbal but also visual and representational. It emphasises the subjective, intimate expressions of comfort, belonging and commitment, which do not always come up in verbal discussions.

In *Part IV: Between the 'Holy' and the 'Global': On Local Knowledge and Spatial Planning* the emphasis moves back to the planning practice. In this part the knowledge identified in the previous parts is challenged with the aim of creating a new language and methodology that merges both local embodied and professional knowledge in the planning practice.

How to read this book

The Global City and the Holy City focuses on three main themes:

1. the three concepts of comfort, belonging and commitment, which are defined as local embodied knowledge;
2. the six categories of environment to which people relate in their narratives of comfort, belonging and commitment (the home, the building, the street, the neighbourhood, the city and city centre, and urban parks and transportation);
3. the two cities that people live in: London and Jerusalem.

In the process of writing the book I had to make a choice on which of the themes to focus on in the analysis and the way to formulate the chapters. I had to decide whether the focus of the book would be on the understanding of the three concepts, or the understanding of the different categories of space or the understanding of what it means to live in Jerusalem and London. Actually the three modes of analysis interlink, and the differences between them are in the emphasis that each makes. In the end I chose to focus on the three concepts of comfort, belonging and commitment as they reflect the idea that I try to promote in the book: that local embodied knowledge matters in planning. Thus each concept is analysed as a chapter in itself, dealing with the meanings and symbolisms of each of the concepts in each category of space in both cities.

However, the reader may find it easier to read 'horizontally', that is, to choose to focus on a specific category of space, for example 'the home', and to explore the meanings of comfort, belonging and commitment at home moving from one chapter to another.

Note

1 The exact term in Hebrew does not have a synonymous word in English. It is both 'calmness' and 'comfort'. I have decided to use the word comfort.

Planning, Knowledge and Diversity in the City

Part I provides the conceptual framework and the theoretical base for the following chapters.

It highlights the book's key issues, such as planning traditions, globalisation and the city, the discourse around knowledge and issues of identity differences and diversities and their implications in city life. It provides not only the conceptual framework but also the contextual framework of the existing planning systems in London and Jerusalem.

Planning traditions, globalisation and the discourse around knowledge:
History, criticism and change

For the 150,000 Muslim Bedouin living in the Negev Desert, South Israel, the different approaches to planning traditions, current globalisation trends and the discourse around knowledge have their expressions in the people's ongoing conflicts with the establishment in Israel. These conflicts focus on two main issues: the landownership dispute and the different perceptions on lifestyle and settlement schemes that the Bedouin and the state bureaucrats have.

The state does not acknowledge the Bedouin land claims and wishes to concentrate them in a small number of towns, each large in population, so that several tribes live in one town. This approach has its base in the professional modernist viewpoint, which perceives modernisation as a desired target that communities wish to achieve. It is also based on the assumption that the planners are the knowers – as agents of professional modernist planning they can assist communities to bridge the gaps between their current level of living and what is perceived as the desired, highly modernised level of living.

However, this is the opposite of what the Bedouin want. Modernisation for them is not a target to be achieved at any price. They are used to living in places that reflect their traditions and norms. This includes, among others, living in a village-like habitation and not in a town, living on lands they claim they own and not on the lands of other tribes, and living in a tribal-based village and not with other tribes as in the towns the state built for them. Their lifestyle expresses their 'local embodied knowledge', a spatial knowledge that derives from their traditions and norms and guides them in their self-planning of their traditional villages.

Today some 70,000 Bedouin live in seven modernised towns, which, as the state claims, provide them with services such as schools and health clinics and infrastructure such as roads, running water, electricity, telephones, etc., but at

the same time these modernised towns are designed on principles that sometimes contradict their traditions. The rest of the population still lives in some
108 traditional, tribal clan-based villages, which are considered illegal in the
eyes of the state, and therefore the state does not provide them with basic
services and infrastructure. These people live without electricity, running
water, schools and health clinics.

This is a clear example of how different perceptions on landownership
patterns and on the connection between cultural norms and the type of habitation needed cause clashes. We can distinguish here between the local knowledge of Bedouin traditions and culture, which is spatial, and the professional
knowledge of the state bureaucrats and planners. This latter knowledge has
been attained in schools of planning and architecture in Israel and abroad,
mostly reflects a modernist approach and has been viewed as a means of
state control over the Bedouin. The fact that the Bedouin towns are still considered inappropriate in the eyes of the Bedouin themselves expresses these
different viewpoints. The Bedouin still prefer to live in their traditional villages, which were constructed on the basis of their cultural norms and daily
practices.

Residential density is an illustration of the differences between the two
types of knowledge. For the Bedouin, human density is a factor affected by
cultural norms of privacy and women's honour. Thus housing density may be
relatively high inside a traditional Bedouin village as it consists of members
of the same affiliation group (tribe, sub-tribe, clan, etc.). This means that the
threat of breaking cultural laws of women's modesty is relatively limited.
But at the same time there must be a great distance between each of the traditional villages in order to minimise the chances of unwanted meetings between
women and men of different tribes. In contrast, the professional planning
knowledge sees high level of residential density as a prerequisite of modernised
living and a concentration of several tribes in one such town is inherent in the
basic idea of modernised planning, providing higher levels of services for large
settlements. For the Bedouin, living in the same town in high density means
increasing threats to breaking codes of privacy and honour. This situation
of high concentration has tremendous effects on Bedouin women's everyday
life because women are more restricted in their freedom to move in what are
termed 'public spaces' (see Fenster 1999a).

Similar contrasts between professional knowledge and local embodied
knowledge had traumatic affects on the Ethiopian Jewish immigrants who
have been in Israel since the 1950s. The 65,000 Ethiopian Jews living in Israel
arrived as a result of dramatic rescue operations prompted by civil war and
famine in Ethiopia, within the context of Israel's national mission to gather
Diaspora Jews from all over the world and bring them to their Jewish homeland. The state of Israel prepared itself for the 'absorption' of the Ethiopian
Jews by formulating two professional Master Plans. These provided policy
guidelines based on previous experiences of absorption of immigrants in Israel,
expressing again a modernist viewpoint and not acknowledging the major

changes the Ethiopian Jews have to undergo, especially in terms of the shift from rural to urban lifestyle, in employment patterns, in family structure and power relations (see Fenster 1998). Here too planning became a means of control, as part of the policy has been to provide the Ethiopian Jews with housing mortgages only in certain towns and cities in order to avoid their concentration in specific places. For them it meant they couldn't choose to live where they wished, but only in designated towns or cities and not necessarily where members of their extended family live. This was also in contrast to their communal way of living in an extended family structure.[1] As with the Bedouin, this modernist view does not allow the Ethiopians to carry on with their traditional rural lifestyle, nor does it provide them with a real option to choose where to live.

We can see how planning traditions in the two examples are based on the modernist professional planning approach, an approach characterised by its centralistic viewpoint and its homogenising perception of communities and individuals. This approach assumes that the professional planning knowledge is scientific, that is, neutral and universal, not taking into account notions of cultural diversity and power relations within communities. Planners thus do not need 'to know a lot' about the people and the communities they plan for, rather they simply use universal principles of their 'professional' knowledge and education. The results of such an approach are usually disastrous for the communities, causing in extreme cases increasing levels of poverty, family violation and drug use. In most such cases the planning approach has been decided on according to principles determined by those in power and according to the 'state's needs', with no attention paid to the needs of the less empowered community members themselves. Clearly indicators such as the type of locality, the size of population and the geographical location of the towns were not evaluated with particular attention to community member's needs and they are therefore worse off in their everyday life. Moreover, with regard to planning for the two communities, professional meetings, plans and documentations were prepared with no involvement of the community members or their representatives. And the end result – the plan – was also unfamiliar to the community members because these were professional documents, formulated in a professional language and accessible only to professional planners or politicians.

The Bedouin and the Ethiopians are two communities that do not perceive city life as their desired lifestyle but who are forced in a way to join the grand modernisation trend of urbanisation. But we could encounter such communities in many global urban spaces. In many cities immigrants' and ethnic minorities' needs and aspirations are becoming less and less relevant to the city's planning and management. As a result, in many global urban spaces the gaps between the 'majority' and the 'minority' are widening, whether these gaps are based on economic, racial, national or ethnic differences, and in fact it is not only cultural minorities who become victims of modernisation but also the 'general public' who are negatively affected by the city's ghettoes, the residential wastelands, the polluted areas, the areas of crime and fear and the changing meanings and relationships between cities and nature. The global city privatised public spheres, turning developers and private entrepreneurs instead of the authorities into the controllers of cities. The authorities for their part are forced to subordinate their plans to the imperatives of the developers (Wilson 1991). The result is that people's needs and aspirations are less expressed in urban planning and management of these cities.

This critical view of the city is not new. In 1961 Jacobs described it in her book *The Death and Life of the Great American Cities*, where she attacked current city planning and rebuilding and the gaps existing between city planning and its users' needs. These gaps are also expressed in Raban's (1998: 156) disappointment of the modernist city:

> Reading Mumford, Howard, Geddes, Corbusier, Park, Weber, I can never manage to believe in their cities or their citizens. It is not that people and social lives in these books are too schematised and sketchy, but that they seem to belong to an entirely different culture from mine; they are Houyhnhnms burdened with a few problematic Yahoos; and the import of a great deal of conventional writing about the city is a complex necessity of persuading apes to turn into horses. The city I live in is one where hobos and loners are thoroughly representative of the place, where superstition thrives, and where people often have to live by reading the signs and surfaces of their environment and interpreting them in terms of private, near-magical codes. Moreover, these people seem to me to be not sports or freaks, but to have responded with instinctive accuracy to the conditions of the city.

Raban writes mostly about London but these contrasts between planners' views and users' views of the city can express the negative feelings that the Bedouin have towards their towns. What Raban refers to is the necessity for a different view or vision of the city, a different planning approach that bridges these gaps between 'writers' and 'readers', between 'planners' and 'users', a planning approach which is sensitive to what Sandercock (1998a) calls: 'the voices of the borderland', a planning approach that relates to the diversities Raban mentions and: 'eases rather than resists the transition to the multicultural cities of the next century, a planning which celebrates and facilitates diversity and difference' (Sandercock 1998a: 119). She calls it 'epistemology of multiplicity', a planning approach that emphasises alternative ways of knowing, knowing both the absolute/objective knowledge and the relative/subjective knowledge: 'we need to acknowledge the many other ways of knowing that exist; to understand their importance to culturally diverse populations; and to discern which ways of knowing are most useful in what circumstances' (Sandercock 1998a: 76).

Following the above critiques, this chapter aims to provide the conceptual background to and literature on the efforts made in the last decades to find better ways of planning, better ways of knowing in planning and better ways of managing our towns and cities, especially in the era of globalisation.

Globalisation, the role of the city and its planning systems

'The city', the symbol of civilisation, has fired the imagination of many writers, poets, philosophers, geographers and planners for many generations. They use various images and metaphors to emphasise its complex structure and functioning: 'The city itself is a text to be read and interpreted,' says Elizabeth Wilson in her book *The Sphinx in the City* (1991). Jonathan Raban (1998: 4) writes: 'the soft city of illusion, myth, aspiration, nightmare, is as real as the hard city one can locate on maps and statistics, in monographs on urban sociology and demography and architecture'. Raban imagines the city not just as a book but also as an encyclopaedia, arguing that this is a useful illustration for planners, philanthropists and journalists who during the nineteenth century resorted to more rigid metaphors of the city. For him this metaphor suggests the special randomness of the city's diversity: 'it hints that, compared with other books or communities, the logic of the city is not of the kind which lends itself to straightforward narration or continuous page to page reading. At the same time it does imply that the city is a repository of knowledge' (88).

The gap between the 'hard' city and the 'soft' city becomes even more evident in the context of the economic global processes of the last few decades. Global cities are defined by Saskia Sassen as: 'sites for the production of specialized services needed by complex organizations for running a spatially dispersed network of factories, offices and service outlets, and the production of financial innovations and the making of markets, both central to the internationalization and expansion of the financial industry' (Sassen 1991: 5). Sassen herself raises a concern regarding these developments: 'What happens to accountability when the leading economic sectors are oriented to a world market and to firms rather than to individuals?' (Sassen 1991: 334). This concern shifts the focus to the 'soft' city and connects us to the role of planning in the age of globalisation as a tool for creating better places to live.

The meanings, roles and perhaps even processes of planning have changed in the era of globalisation. This era resulted in new organisational landscapes, which emerged in the mid-1980s and are dominated by accountants, business or corporate values and technicist procedures and discourses (Imrie, 1999). Historically, planning schemes were always the domain of the nation–state bureaucrats, making sure that urban planning expressed national goals. In the last few years the growing importance of the global economies and the weakening of the nation–state apparatus meant other forms of local government taking responsibility on planning matters. These changes are connected to prioritising efficiency over values, profit over welfare and private interests over public interests. In planning terms efficiency, profit and private interests mean the speeding up of turnaround of planning applications, the completion of local plan preparation, the facilitation of development objectives and the streamlining of procedures such as public consultation, which potentially slow down the plan-making process. Moreover, planners are subjected to a managerialist culture within local government, which seeks to divide and disperse functions (Imrie 1999). Clark and Stewart (1997) argue that this trend is actually linked to the emergence of a government culture of target setting, performance management and the building of corporate commitment, which is almost a must for surviving in such a competitive environment. A reflection of these trends is expressed in Kanter's (1995) argument that for the city to become 'globally excellent' it needs to be managed and not planned. It needs to be managed as a coalition of partners, which can tackle urban problems, by being managed as a business. This means making local government 'businesslike' and marketing public services provision including planning (Imrie 1999). In these new situations the concerns expressed by Sassen become real and explicit, and the question as to what will be the role of the planners and for whom to plan becomes crucial. It emphasises the fact that not only cultural minorities such as the Bedouin and the Ethiopians in Israel are affected by processes of modernisation and globalisation but the 'general public', and especially the 'have nots' and the 'less powerful', are becoming more and more marginalised in such urban spaces.

This rather new situation of globalisation and its effects on city planning and management leaves radical and insurgent planners with feelings of unease as to how the 'voices of the borderland' would be raised in such an efficient and 'consumption oriented' planning system. In order to tackle these concerns we must look first at the history of planning traditions since the end of the eighteenth century. This I do in the next section with particular reference to the notion of knowledge and its role in planning traditions.

The history of planning traditions – the discourse of universal vs local knowledge

The meanings of universal and local knowledge

'I draw a distinction between *ideas* which can be put into the world in a disembodied way, variously spoken, written down and electronically communicated,' writes Ma Reha 'and *knowledge* which must be embodied and cannot exist outside bodies . . . *knowledge* is within us as human beings and it arises from ideas, perceptions, mental formations, information and feelings that come into our bodies through our senses or are created internally by our mind' (Ma Reha 1998: 2). *Ideas* are in fact representations of the universal or absolute knowledge, while *knowledge* is reflected in what is termed 'local knowledge', embodied in people's everyday life experiences.

Are the dichotomies of the two types of knowledge indeed so separate in reality? Is it an either/or situation or perhaps in many cases, including cases analysed later in this book, that such dichotomies are not so explicit after all?

In what follows the two concepts will be analysed as two separate entities. Later on we will see that sometimes there are meeting points between the two types of knowledge depending on the type of planning approach and the role of the planner in the planning process.

Let us start with the distinction between universal and local knowledge as it appears in the literature. As Sandra Harding writes (1996: 444):

> all knowledge is local . . . modern northern scientific knowledge no less than the 'ethno sciences' of the other cultures . . . the major difference between the two types of knowledge is their power. Obviously not all such local knowledge systems (LKS) are equally powerful, some can explain a great deal more than others because, for example, they have access to or control more of nature or social relations . . . We could think of these as positional rather than substantive epistemological and scientific resources to mark how it is a position in power relations with respect to other cultures that generate such resources.

Harding argues that in fact the two types of knowledge are local, the white western knowledge as much as the local ethnocentric knowledge. The distinction between them, she argues, is about power relations, the white superiority and political power, which has created the colonialism and imperialism in the last centuries, makes one type of knowledge considered as 'professional', 'universal', and the other type as merely 'local'. Moreover, post-colonial critics point out that the epistemological dominance of northern thought has not in fact been established through the internal 'properties' of European scientific and epistemological methods but has been developed through European expansion's power to test scientific hypotheses across extremely diverse local natural conditions or, in Harding's words: 'how European expansion turned the world into a laboratory for European sciences, in the process of destroying the scientific and technological traditions of other cultures' (1996: 440).

This strong statement places a big question mark around the legitimate superiority of the 'white knowledge' of modernity in planning. Taking this point, let us go back to the discussion on the 'professional' and 'local' knowledge in planning. Professional knowledge in planning can be defined using Ma Reha's terms, as 'academically generated ideas' that were produced in western universities and schools of planning from the mid-nineteenth century by various disciplines, mainly in Britain and the United States, and were spread all over the world as the ultimate authority of scientific knowledge in planning and thus became the knowledge appropriate to all places in the world. 'The central task of the British Empire,' writes Ma Reha, 'was the preparedness of Oxbridge graduates to [go to] the colonies and establish universities and other colonial structures, there by spreading a new global awareness based on [the] British way of doing things' (1998: 3). In planning it relates to graduates of schools of planning and architecture in Europe and the United States who went back to their homelands not only to establish universities but to reshape cities, recreate urban spaces and relocate urban services, usually under the aegis of European planners and architects who looked on local traditions, norms and values as old-fashioned and irrelevant. Reshaping of cityscapes was then made according to the white, 'Oxbridge' knowledge that these graduates possessed. It was only in the early 1960s that theorists and critics began to have doubts about whether the importation of Northern models of planning and development was effective in enabling local societies to catch up with Northern standards of living. This is clearly expressed in what happened in the 1960s in planning traditions, as will be discussed below, where the focus of planning and development shifted from how to modernise urban spaces into how to approach its users. This shift would be more suitable to communities such as the Bedouin and the Ethiopians in Israel in their struggle with the effects of urbanisation on their lifestyle.

In reviewing the different planning approaches that have developed from the eighteenth century I have used several categorisations from the literature: the classification presented by Safier (1990)[2] and Moser (1993), which looks at three main planning traditions: the physical planning traditions, the socio-economic and the transformative planning traditions. This classification is compatible with the categorisation Sandercock made (1998a), looking at the rational comprehensive model and alternative planning approaches. Fainstein (2000) chooses to emphasise the changes in concentrating on three planning modes; the rational model, the communicative model and the just-city model. In what follows I will present another classification of planning approaches highlighting how each perceives and illustrates the tensions between 'professional' and 'local' knowledge.

'The professional planning knowledge': the procedural, modernist, rational comprehensive planning approach

The rational model

The rational comprehensive approach (RCA) is considered as the ultimate 'scientific method', a western expression of modernity in planning. It represents procedural planning tradition and is considered a formal, top-down process emphasising the need for intervention by the state in the market and the belief in greater rationality in public policy decision making (Sandercock 1998a). It illustrates the belief of the era after the Second World War that technology and social science could make the world work better and that planning could be an important tool for social progress. It provided the 'meta theory for planning activity' (Fainstein 2000). Theorists from Simon (bounded rationality) to Lindenblum (incremental decision making) to Etzioni (mixed scanning) shared common faith in instrumental rationality and in its ability to make the world better. RCA has been used primarily for forecasting impacts and for programme evaluation (Fainstein 2000). Beauregard (1987: 367 in Fainstein 2000) mentions that: 'in its fullest development, the Rational Model had neither subject nor object. It ignored the nature of the agents who carried out planning and was indifferent to the object of their efforts (i.e., the built environment).'

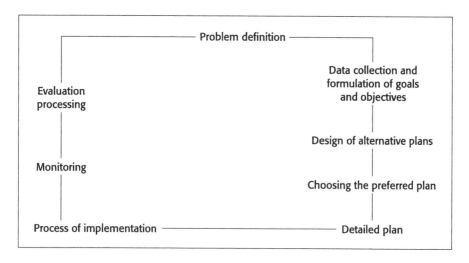

Figure 2.1 The rational comprehensive model of planning
(Source: based on Moser 1993)

As seen later in the chapter, this type of criticism opened up new directions in planning theory.

Figure 2.1 presents the different stages of the rational comprehensive model.

The underlying assumption of this model is that most of its stages are based on the planner's knowledge. In this model the planner is the 'handmaiden to power' (Sandercock 1998a); he is 'the knower', relying strictly on 'his' professional expertise and objectivity to do what is best for an undifferentiated public. The notion of 'public' is never critically examined within planning. Differences of class, race, gender or culture were not considered relevant. It assumes a state whose structure is neutral with respect to questions of gender, race, ethnicity, culture and sexual preference. This type of western modernist theory views the process and direction of change as predetermined, while notions of empowerment and forms of social and environmental relations are ignored. Furthermore, modernised planning involves an autonomous process of change rather than being a product of the integration of pre-modernised cultural codes, norms, values, social relations and environmental attitudes. It suggests a 'scientific' method for planning that is quantitative and accurate and thus offers the illusion of objectivity and certainty.

The rational comprehensive approach produced a variety of planning traditions. Safier (1990) terms them the classical/physical traditions. Classical/physical traditions are concerned with physical and spatial problems of city, and regional growth with the 'blueprint' approach as one of its methodologies, and from the 1950s onwards the rational comprehensive approach became its popular methodology. Table 2.1 provides details on the four planning approaches included in this tradition: urban design, town planning, regional planning and transport planning. Its methodology clearly emphasises

Table 2.1   Classical planning traditions – emphasising physical aspects

	Urban design	Town planning	Regional planning	Transport planning
Period	1890–1920	1840–1914	1930s	1950s
Origin	Europe, US	Britain	US & USSR	US
Ideologists	Le Courbusier	Howard	Geddes, Mumford	Kaim, Mayer
Focus	Built form	Urban land	Metropolitan special view	Movement
Knowledge base	Architecture	Estate management	Geography	Engineering
Objective	Function	Order	Balance	Efficiency

strong knowledge–power dimensions in planning because it assumes that state bureaucrats can be and are the ultimate authority for translating the information about and understanding economic, social and environmental needs of the population in physical and spatial forms. Procedural planning assumes spaces as absolute and perceives its physical aspects rather than its social and cultural constructions. These approaches can be looked at as expressions of 'Oxbridge white academic ideas in planning', which were imported as the ultimate science for the rest of the world.

Within two decades this tradition faced challenges from within as well as from without (Innes 1995). It has been criticised as insensitive to the complexities of social change and to the metaphorical, sometimes hidden, meanings attributed to space. In addition, the age of the global economy has exposed other challenges that the rational comprehensive planning model faces, such as the increasing involvement of local non-governmental organisations and others that advocate needs and desires based on local knowledge. Also, some of the world cities act independently to the extent that they become 'city–states' departing from the nation–state political structures so that the role of central planning and the RCA is weakening. Perhaps a much more complicated question that the rational comprehensive planning model does not challenge is how cities and regions that, according to neo-liberalism, are striving to make themselves more economically competitive can also become 'better' places to live within local contexts. Neo-Liberalism puts cities in a highly complicated position. Cities are becoming both attractive and repulsive places in which to live and work. One of the challenges of planning and management in the twenty-first century is therefore how to make cities a place like home for their various communities.

Despite the criticism, it must be emphasised that the major advantage of the rational comprehensive model is in the reformulation of the 'know-how', a procedure or method of planning that is rational, clear, simple and coherent. This model also offers decision rules that are logical and clear and which allow planners to study alternatives and consequences (Sandercock 1998a). Because of that and the powerful hegemony of the white knowledge this model is still dominant in the field of planning.

'The local based body of knowledge': alternative planning traditions

A series of alternative planning approaches arose from the 1960s onwards trying to accomplish what the rational comprehensive model is missing in terms of more emphasis on local knowledge in the different planning stages. These alternative planning traditions focus on the people who are supposed to be affected by the planning procedures. Their rationale is highly connected to the aim of creating a different body of knowledge or perhaps ideas needed for planning. While Sandercock (1998a) includes within this division some of the more famous planning traditions such as advocacy planning, the radical political and economy approach, the equity planning model, and so on, Fainstein (2000) only concentrates on the communicative model as representing this new line of thinking. Here is a brief review of each of these traditions with reference to the tensions between professional/local knowledge that each represents.

Advocacy planning approach

This was developed by Davidoff (1965), who suggested incorporating several plans as part of the planning procedure rather than one master plan. Each of these plans represents the interests of the group that the planner represents. The public, argues Davidoff, should be more involved in the planning process and the planner should serve as advocate of the different groups involved. Here local knowledge is the core of the planning process but the planner is still the ultimate knower, representing the community in negotiations with the authorities.

The radical political and economy planning approach

This approach was formulated mainly by Harvey (1973) and Castells (1976), who criticise rational comprehensive planning from a Marxist perspective as a function of a capitalist state. Planning in capitalist society is both necessary and impossible (Castells 1976). Urban planning, they argue, is mostly based on zoning and other regulatory mechanisms, the role of which is to bring a balance between competing factions of capital and between capital and citizens through a mixture of repression, cooptation and integration (Harvey 1973). This model however, has not provided an alternative definition and methodology of planning.

The equity planning model

This model focuses on urban inequalities, and the close working of planners and politicians in order to try and bridge the gaps of these inequalities. The planner, however, is still the key actor, the 'knower'.

Table 2.2 Applied planning traditions emphasising socio-economic aspects

	Economic planning	Project planning	Corporate planning	Social planning
Period	1970s	1970s	1960–1970	1945–1975
Origin	USSR	World Bank		UK (colonial period)
Focus	Resources	Investments	Organisation	Communities
Knowledge base	Economic	Finance	Management	Social work Sociology
Objective	Growth	Economic efficiency	Financial integration	Welfare participation

The radical planning model

This model of the 1980s represents a mixture of advocacy, feminist critique and the civil rights movement and has been engaged with multiple critical discourses about social transformation. It emerged from the critique of existing unequal relations and distributions of power, opportunity and resources. The goal of this model is to empower the disempowered. It recognises the heterogeneity of communities and views the role of the planners as alleviating existing inequalities. Obviously this model engages local knowledge in the planning process as it is based on advocating and promoting people's needs.

Safier (1990) and Moser (1993) include in these traditions what they term 'applied' and 'transformative' planning traditions. The 'applied' planning traditions shifted their concern from spatial and physical domains to economic and social growth at both project and corporate levels. Applied traditions are also identified with a rational comprehensive methodology but here we notice a shift in the focus of planning traditions from the physical space to human activities and the beginning of a shift from professional ideas to local knowledge. Table 2.2 presents the four types of planning traditions included in this category: economic planning, project planning, corporate planning and social planning.

Transformative planning traditions are considered the alternative methodologies to the rational comprehensive model. These methodologies allow the 'voices from below' or the voices of the 'other' to be heard and the 'power game' between individuals/groups and the elite takes a more dominant and explicit role. 'Transformative' traditions require much more transformative procedures than the other pervasive methodologies. These traditions include development, cultural, environmental and gender planning and are obviously more sensitive to diversity and human rights issues and as such are more focused on local knowledge. Table 2.3 represents the categories included in this tradition.

Table 2.3   Transformative planning traditions emphasising human aspects

	Development planning	Gender planning	Culturally sensitive planning	Environmental planning
Period	1960s	1980s	1970s	1970s
Focus	Needs	Gender	Culture	Environment
Knowledge base	Development studies	Feminist studies	Historical experience	Environmental studies
Objective	Development	Emancipation	Identity	Adaptation

Modernist planning, alternative planning approaches and feminism

Feminists criticise modernist planning too and see such a model as a male bastion (Sandercock and Forsyth 1996). Modernist components such as residential zoning, which emphasise the division of home from work, or the distinction between downtown and neighbourhood, are seen as reiterating and reinforcing the familiar distinctions between the male and female spheres (Fainstein 1996). Still, there are diverse opinions as to where the analysis of gender is relevant, especially regarding the economic status of women; their location and movement through space, the connection between capitalist production and patriarchal relationships, the connection between 'public' and 'private' life; how women know about the world and how much they learn to define their needs, and finally what forms of communication women feel most comfortable with (Sandercock and Forsyth 1996). What is the significance of feminist contributions to planning theory? These contributions serve to highlight the control over women in planning policy and the resultant negative effects on them, and lately in making the connection between women's human rights and town planning. The latter refers to the extent to which initiatives to meet women's needs in the built environment are successful as well as the extent to which women's involvement in decision making within the local authority serve as a vehicle to meet women's needs (Little 1999).

The communicative action model

This model draws on two philosophical approaches – US pragmatism and the theory of communicative rationality as worked out by Jurgen Habermas (Fainstein 2000). While the first approach tends to empiricism the latter starts with abstract proposition. The two rather contrasting resources provide a guide for action to planners. Thus within the communicative model the planner's primary role is to listen to people's stories and assist in forging a consensus among different viewpoints (Fainstein 2000). Moreover it emphasises a mutual learning process between the planners and the 'customers'. Friedman (1973) calls it 'transactive' planning as a life of dialogues emphasising human

Table 2.4 A dichotomy of tensions

Professional knowledge x	Local knowledge x
x -Rational comprehensive -Bounded rationality -Incremental decision making	x -Radical planning - Advocacy planning -Communicative -Insurgent planning

worth and reciprocity rather then providing technocratic leadership. Forester (1989), Healey (1992, 1997) and Innes (1995) developed this approach precisely in order to create the link between the professional knowledge and the local knowledge. Healey in particular emphasises how all forms of knowledge are socially constructed and that knowledge may take many different forms including storytelling and subjective statements. Local knowledge is produced socially and is being developed through interactions and experiences of individuals. In this respect the communicative action model together with some of the ideas of the radical planning model aspire to develop a new 'epistemology of multiplicity' (Sandercock 1998a), which perhaps is getting closer to what Ma Reha terms as 'knowledge'. Its basic premise is the acknowledgement of 'local knowledge', that of hearing and listening to the 'voices of the borderline', that is, women and people of colour, people of different national, ethnic and cultural origins in the planning process. For initiating this approach Sandercock mentions 'alternative methods of knowing, learning, discovering, including traditional ethnic or culturally specific modes: from talk to storytelling, to blues to rap, poetry and song; and visual representations, from cartoons to murals, paintings to quilts' (1998a: 121). This model relies more on qualitative, interpretative inquiry than on logical deductive analysis, and it seeks to understand the unique and the contextual, rather than make general propositions about a mythical, abstract planner.

To sum up, the tensions between professional knowledge on the one extreme and local knowledge on the other can be illustrated as a dichotomy as shown in Table 2.4. Its dichotomous nature is used here to clarify the differences between the two channels of knowledge although it is well known that in reality the two usually intermingle.

Local knowledge and the rational comprehensive approach – are the two compatible?

One way of intermingling these two channels of knowledge is by using this methodology to incorporate 'local based knowledge' into the planning practice. RCA can be a useful tool for integrating *planner's* and *user's* knowledge in each stage of the planning process. This view actually makes a distinction between the *how*, that is, *the method* that this model suggests for planning procedure and the *what*, that is, *the content* or the type of knowledge that each of its stages uses for planning. Since the *how* is so far the most elaborate procedure for planning it can serve alternative planning approaches with a different 'content' in each of the stages involving different types of knowledge and a different balance between professional and local knowledge. Obviously this means that the role and the dominancy of the planner in the planning process is changing as well.

The role of the planner in the different planning traditions

The role of the planner varies according to the social and power relations embedded in each planning tradition presented above. As mentioned, according to the rational comprehensive approach, the planner is seen as the knower, the ultimate authority of knowledge, the isolated professional who carries out work with minor contact with the customers. In contrast to this view, alternative planning approaches place a greater role for the planner as a community transformer. Sandercock (1999) argues that radical planners, for example, should be trained professionals as change agents; their role is central in the social transformation process that radical planning is aimed to bring. This role can also create a paradox. For example, sometimes those committed professionals who work with mobilised communities find themselves 'technicians' in the service of the 'client', the mobilised communities, with no option (or wish) to impose their values on that community (Lane 1999). Another paradox is that sometimes committed professionals find themselves working for mobilised communities in the traditional rational comprehensive approach doing battles with other planners of the state. Here the professional knowledge they are expected to provide is merely the technical knowledge of how to use the planning language of the state, which will enable them to put community ideas into the recognised 'text' – the plan.

These are not the only dilemmas that planners experiences. Harvey (1999: 272) argues that: 'planning is about confronting the dialectic in its "either/or" rather than its transcendent "both/and" form. It always has its existential moments'. One such either/or situation is what he calls: 'the personal is political (and the planner is a person)'. That is to say that the planner, like any other, is always a political person and in reflecting what planners do it is vital to leave a space for 'the private' and 'the personal' of the planner's identity. Furthermore: 'the planner can not change the world without changing herself or himself.' Thus the negotiation that always lies at the basis of the planning practice is between political persons seeking to change each other and the world as well as themselves.

These dilemmas and different attitudes to their roles are analysed in Chapter 10, where various planners relate their roles in the planning processes they were involved in.

Another dilemma inherent in the planner's role is the positionality of the planner as to what constitutes social transformation and what are the planner's abilities to move from where they are in their positional situation to somewhere else in order to create change. This is especially relevant in cases when the planner comes from a different background to the communities they work for. The dilemma then is what are the directions of change and who makes the decision of which changes are positive or negative to some members of the community. This kind of dilemma can be termed: 'the dialectics between particularities and universalities' and is elaborated in the last section of this chapter.

Returning to the notion of knowledge, perhaps the most dominant dilemma for planners is how to balance professional ideas and knowledge with local knowledge in the context of power and control. The next stage in the discussion is the contextualisation of current planning traditions with theories of power and knowledge. The rest of the chapter aims to do this.

Planning as control by means of manipulating 'knowledge' – the creation of the 'misuse of knowledge'

In the last section I reviewed the development of planning traditions emphasising the tensions between professional and local knowledge. In this section I move a step forward and present the notion of 'misuse of knowledge' to emphasise how, by looking at knowledge as universal and neutral, as modernist planning does, it has the potential to become a tool of control. Following this argument, I analyse the creation of three different types of misuse of knowledge: the professional, the local symbolic and the local gendered misuse of knowledge. To illustrate the use of these three concepts I analyse them within the context of the two case studies presented at the beginning of the chapter, the Bedouin and the Ethiopian communities in Israel, but this time with an emphasis on gender issues.

First, let us define the term 'misuse of knowledge'. Misuse of knowledge can be expressed in two ways. From the side of the planners, the term emphasises planners' tendency to ignore existing sources of knowledge, especially local symbolic and local gendered knowledge, which is so essential for formulating efficient and adequate plans in particular for non-Western societies. From the users' side, the misuse of knowledge is manifested in the lack of access of the 'have-nots', the 'others', the powerless to existing sources of knowledge, usually technical-professional sources that are part of the planning process. These sources are usually kept in the hands of those in power, that is, authorities, dominant groups in communities (such as, men, dominant ethnic groups, etc). This situation of 'misuse of knowledge' is different from manipulation of knowledge in that it relates not only to the actual professional manipulation of knowledge by planners that sometimes occurs in the process of planning[3] but also to the lack of access to sources of knowledge such as plans, development proposals and programmes. Misuse of knowledge can thus be perceived as one of the expressions of planning as a mechanism of control.

Planning as a control system

Planning creates and reflects power relations in two ways: by controlling material resources in the environment, such as land, water, oil, minerals, etc., and by controlling metaphorical or local symbolic resources such as knowledge. Giddens (1981) makes a similar distinction between two types of resource that are controlled by the state: authoritative resources and material resources. Lack of authoritative resources, he argues, involves primarily the retention and control of information or knowledge. Since space is such an obvious and fundamental aspect for any form of communal life and of any exercise of power (Foucault 1980), the planning of space becomes the 'natural' field of expression of power relations. Indeed Huxley (1994) argues that planning in its control of space can be seen as part of what she calls 'the normalization generated through the exercise of power'. Following this line of thought I defined 'planning' elsewhere (2002) in a rather Foucaultian spirit as the arrangement of space according to principles and goals determined by those in power. Some would argue that plans control all citizens' or community members' 'free will'. However, if those plans are spatial translations of goals of a particular group (men, ethnic, class, race, cultural majority) they obviously serve the interests of this group, which in many cases contradict principles or goals of other groups.

There are other perspectives in the literature that explore the connection between planning, control and social structures. Forester (1989) emphasises the relationships between planners and those in power, arguing that if planners ignore those in power they assure their own powerlessness and of course the lack of power of their plans. Yiftachel (1995) discusses the conversion of planning from a progressive tool of reform to an instrument of control and repression when one ethnic group becomes dominant and uses the state's apparatus (including planning) to maintain and strengthen its domination. Yiftachel analyses planning as control in three principal domains: territorial, procedural and socio-economic, and shows how these three are used to control the Arab-Palestinians in Israel. The contribution of Fainstein (1996) to the discussion of planning as control is by highlighting the feminist perspective, emphasising the male dominancy with urban planning, which neglects the interests of women, especially in economic development and land use planning. And finally, Harvey (1985, 1996) emphasises how changing global economies and the decreasing dominancy of the state brings changes in power relations within planning schemes, making more room for 'bottom-up' initiatives to emerge and allow local forces and local knowledge to become part of the planning process. This means that global structures and their effects on planning allow more room for local knowledge to be heard and to affect planning.

What are the expressions of control in planning? In the ways and means by which planning procedures define and allocate land uses, that is, the restrictions of the use of land to what has been decided upon by those in power. This is mainly expressed in the articulation of zoning principles, which are formulated to protect the interests of the 'public'. Zoning is used to posit homogenous areas that exclude the other: 'it is a part of the state's involvement in the creation and control of citizens/subjects and the discipline of bodies' (Huxley 1994: 8). But zoning is only one of the state's means to exclude others. Policies of economic oppression, land confiscation and political control are other means to exclude communities. Zoning regulations, argues Huxley, used to posit homogenous areas, to leave them 'purified' from the different, the minority. It is part of the state's involvement in the creation of a means of control of citizens (Huxley 1994).

While some view the 1950s as the 'golden age' of planning, as a positive and central agent of progressive transformation, others highlight the increasing state power behind planning that sometimes causes inequality and the deepening of intergroup inequalities (Yiftachel 2000). One thus has to identify the fine line where planning moves from reform to becoming a means of control, and where the difference between 'reasonable exigencies of social order' and sinister expressions of repression, exploitation and oppression occur within planning schemes (Yiftachel 1995). Following this distinction it is also important to note that planning as control usually has a stronger impact on the powerless members of a community (such as women and ethnic minorities) than on the elite who are more represented by those in power. However, as mentioned earlier, global changes and the changing role of the state in planning and development leaves more room for voices of resistance that are beginning to be heard, both via NGOs and other forms of organisation or via 'grassroots' protest and resistance groups emerging from those communities that would be negatively affected by planning and development.

I move now to discuss the control of knowledge. The first question to ask is, which 'knowledge'? The 'everyday knowledge' of common people (Innes 1989 in Healey 1992) or 'politician knowledge', the knowledge of the 'elite' (Mazza 1986 in Healey 1992). There are obviously many ways of defining knowledge. I argue here that knowledge in principle is situated, that is, 'partial': 'all knowledge is produced in specific circumstances and those circumstances shape it in some way' (Rose 1997: 305). This means that the sort of knowledge created depends on who its makers are. Knowledge is thus not universal and neutral; it is reflexive and positioned. This new feminist, post-Marxist and post-colonial notion of knowledge has begun to infiltrate the field of planning only recently. Healey (1992) mentions that the systematised, rationally grounded knowledge is now understood to be only one among several knowledge forms. Healey herself acknowledges that any form of knowledge is infused with ideological and political practices that protect the powerful and confuse the powerless.

Planning, argues Healey, must find ways to challenge what Forester refers to as 'misinformation' and I call 'misuse of knowledge'. Three forms of misuse of knowledge are discussed here: professional, local symbolic and local gendered.

The misuse of professional knowledge

'Knowledge' in the professional sense means the knowledge of the profession of planning. Planners and mapmakers were from early days part of the elite. The knowledge of mapmaking, like the knowledge of plan making, was one of the specialised intellectual weapons by which power could be gained, administered and legitimised through history. This knowledge was concentrated in relatively few hands, and maps were associated with the religious and political elite. This is expressed in several periods of history: early Egypt, medieval Europe, Greece and Rome, and in the Mediterranean world during the late Middle Ages (Harley 1988). The dominant group, in our case politicians and planners, decide upon principles of, for example, zoning because they possess either the power or the professional knowledge – positioned and situated – and this knowledge provides them with the power that, in turn, produces knowledge, again situated and positioned (Foucault 1980). This interpretation of the notion of knowledge reflects Foucault's thinking about the relationships between discourse, power, control and the manipulation of knowledge, and ultimately of society (Moore 1996). Foucault views texts and maps as a means of concealment and of deliberate control and as instruments of power/knowledge. Planning maps and texts, such as documentation and guidelines, are illustrations of how knowledge or misuse of knowledge becomes control, as those maps and texts consist of complicated language, known only to professional planners, and the hidden agendas underlying their creation form a 'knowledge' shared only by exclusive people and not by those who are or might be affected by them. Similarly, Forester (1989) calls on us to avoid what he terms 'misinformation' in planning. He emphasises the importance of having 'information' in the planning process and categorises several kinds of information according to the planning tradition adopted (see also Sandercock 1998a).

The misuse of local symbolic knowledge

Local symbolic knowledge is based on the concept of understanding and listening to the others, the marginalised, the have-nots. It is local knowledge, which looks at how people perceive, use and relate to space. It is local symbolic because it derives out of people's culture, norms, habits, fears, emotions; components that create the 'local symbolic' viewing of a 'physical' space. It is not a visible knowledge but something that needs deeper understanding and listening. It is also embodied knowledge in that it is a knowledge we experience through our body experiences. This local symbolic knowledge affects the way people perceive and use spaces and influences their environmental needs. It becomes relevant when a planning intervention is going to be made, in order to make sure that this intervention meets the community's needs. Local symbolic or local knowledge provides the social and the cultural mapping of the physical space, a knowledge needed as a basis for rearrangement of space. This type of knowledge:

> emphasizes listening and interpreting, developing skills that are sensitive to and able to pick up on everyday ways of knowing. It suggests an entirely different practice in which we respect class, gender and ethnic differences in ways of knowing, and actively try to learn and practice those ways in order to foster a more inclusive and democratic planning. (Sandercock 1998a: 121)

Misuse of local symbolic knowledge takes place when planners ignore it, usually because it is not considered relevant to the professional, knowledge – based planning process. Plans based on misusing local symbolic knowledge become irrelevant, inadequate and inappropriate for communities, especially for those coming from non-Western cultural backgrounds. There is another meaning to the misuse of local symbolic knowledge. It can also occur when there is 'a colonization of the Other' (Sibley 1998) that is, 'an appropriation of the minority's knowledge which is not denigrated but becomes a part of the currency of academic discourse' (p. 98). Since this 'private', 'indigenous' knowledge, argues Sibley, is now publicised it can be misused by those intending to control the 'Other'. Here we go back to the knowledge/power relations but from a different perspective. Sibley argues that for this knowledge to be helpful to the 'Other', the 'Other' has to have some control over it, otherwise it increases oppression and discrimination.

The misuse of local gendered knowledge

The third use of 'knowledge' is in the context of gender relations. Spain (1993) elaborates on this idea. Local gendered spaces in non-industrial societies, she argues, remove women from knowledge. For example, in the Mongolian Ger, books are kept in the male part of the traditional tent. Because books contain religious and historical knowledge Mongolian women are forbidden to read them or even look at them because of the fear that women destroy social order. Knowledge, says Moore (1996), like language, is thought to be a male attribute. Access to knowledge, like the control of language, is a prerequisite for full adult male status. Women do not possess social (male) knowledge and therefore are excluded from the exercise of power. Cultural codes dictate the local gendered access to knowledge, that is, cultural norms dictate whether women have or do not have access to sources of knowledge. In most cultures the lack of access to sources of knowledge is spatial. It serves to perpetuate women's subordination and the control of men over them.

As Rose (1997) argues, women are usually the objects of knowledge, not the makers of it. One way of exercising power and control over women is by manipulating meanings of spaces (forbidden, private) in ways that prevent access of women to resources of knowledge which exist in 'forbidden spaces' or the public sphere, such as gaining access to education on its different levels (from literacy to professional training), being exposed to and being able to read newspapers, watching films and watching television, participating in public life activities such as cultural events and political events and many more. This is in many ways similar to the way states manipulate knowledge in planning and control of space because of the fear that minorities would destroy the social order.

The misuse of local gendered knowledge means not relating to the different and unequal access to sources of knowledge by males and females in the community. The opposite situation, that is the appropriate use of local gendered knowledge, would then mean finding out whether these unequal structures of access to knowledge could be altered as part of the planning process so that spaces become equally accessible for both men and women.

In analysing the two case studies presented at the beginning of the chapter I will illustrate how the misuse of local gendered knowledge increases control over women in the planning process for both Bedouin and Ethiopian communities. These cases emphasise how, by enforcing the rational comprehensive–modernist planning approach, many principles of justice, equality and human rights are abused (see Fenster 1997, 1998, 1999b,c).

Planning in practice – the misuse of knowledge as a means of control over Bedouin and Ethiopian women in Israel

Before analysing the meanings of misuse of knowledge in the case of the Bedouin and the Ethiopian communities in Israel let us briefly look at the situation and status of women in Israel within the context of planning.

Women and planning in Israel – a brief background

The state of Israel was established in 1948 as a Jewish state, a home for the Jewish people from the Diaspora, some of whom were Holocaust survivors. The arrival of Jewish communities from countries of different cultural backgrounds created a very heterogeneous society, with different and sometimes contradictory sets of norms and values for each community, especially regarding those of women's role in the family, women's modesty, equality within the family unit and so on.

By definition the state categorises all of its non-Jewish citizens (some 20 per cent) as minorities.[4] Being a Jewish state where spaces are strongly classified and have strong emotional symbols means that one of the important priorities of planning is control (Yiftachel 1995). This means that cultural minorities in Israel face constant political, social and cultural discrimination on the basis of their minority religious position.

In comparison with Western countries Israel has a very high population density: the total per sq km in 1998 was 272.8. Over 90 per cent of residents live in urban areas, about 70 per cent of these live in apartments. Of the total population 73 per cent live in owner–occupied dwellings (Central Bureau of Statistics 2000). Israel has a young population with a medium age of 26.7. Culturally, it is a family-oriented society. All ethnic groups in Israel, both Jews and non-Jews, place a great emphasis on family, as evidenced by the high fertility rate. Another statistic, which reflects the country's family-centred nature, is the low proportion of unmarried Israelis, just 4.3 per cent of men and 5.6 per cent of women in the age group 40–44 among the Jewish population (the figures decrease for non-Jews), and the low divorce rate of 1.6 per thousand in the Jewish population, and 0.8 per thousand among Muslims. Of Israeli households 25 per cent are headed by women (Statistical Yearbook 2000).

Women constitute 50.4 per cent of the population in Israel and most face considerable control, first as part of a socially, economically and politically misrepresented and disadvantaged group, and second, some of them also face control and subordination as a result of their own cultural norms and values. For Jewish women this subordination is mainly expressed in Jewish family laws of marriage and divorce, which discriminate between men and women. Jewish religious women are also affected by norms of modesty that restrict their movement in public spaces. The subordination of Arab–Palestinian women is more explicit: first as members of the discriminated Arab–Palestinian community and second as members of their own culture, which by norms of honour and modesty restrict their everyday life.

Israel has a high percentage of women in the workforce, but this applies mostly to the Jewish population. In general, 51.4 per cent of the total employees in 1999 were women. Among Jewish women the proportion is 52.5 per cent, among Arab women 15 per cent and among Bedouin women only 5 per cent. These differences are a result of varying work opportunities; opportunities are much higher for Jewish women due to policies of economic discrimination but also because of different cultural norms, which allow Jewish women to work outside their home to a greater extent than Muslim and Arab/Christian cultural norms allow their women.

Both the Jewish and Islamic religions are similar in their patriarchal perceptions of women's 'fragile' nature, which is used to dictate a strong control over their modesty and honour. The result is that according to both religions women can not be equal, family laws subordinate them and their place is considered the private sphere of domesticity. Muslim women in Israel sometimes face stronger restrictions because of the Shari'a family laws, which aim to keep women's honour and modesty by strict laws of marriage and divorce. Moreover, since Muslim sexuality is actually territorial (Mernissi 1987) rules of modesty, which restrict women's movement in 'public' and allocate space to each sex, are very clear-cut and well kept, especially among the Bedouin. Muslim women's 'burden of representation' in Israel and in many Muslim countries is sometimes fatal. Women have been murdered by their male relatives because of behaviour that supposedly brought 'shame' on their family and community. Each murder raises public discussions as to the official state policy and the role of the police in preventing such crimes (Rabinowitz 1995). The murder of women is however not exclusive to the Arab–Palestinian population. In 1997 husbands or partners murdered some tens of Jewish women, and the police and the justice system have not succeeded in reducing this violence.

These issues highlight the importance of cultural values as points of reference in the discussion of women's access to sources of knowledge in public spaces. Cultural values refer not only to religious laws and norms but also to the centrality of family life in Israel, the high reproduction rate and the high proportion of Jewish women in the workforce as opposed to the low rate of Arab and Bedouin women. For women to live in a family-oriented society, with the traditional women's roles expected of them on the one hand, and being part of a modernised society with increasing work and career opportunities on the other, is a highly problematic situation.

The planning system in Israel is highly centralised and homogenised. It fails to examine and incorporate aspects of social, cultural, economic and gender diversity. Plans in Israel are still planned for the 'average person' who is: 'a man, Jewish, healthy, 35 years old, married with two children, secular and middle class'. This probably represents not more than 5 per cent of the entire population (Churchman et al. 1996). In the light of the above, the Israeli Women's Network (Churchman et al. 1996) published a shadow report for the 1996 Habitat International Conference in Istanbul. The report sets out some of the principles of women's roles and needs that need to be considered in urban planning. Some of the principles discussed include the geographic location of settlements, the type of locality and the size of population, the frequency of public transportation services, zoning policies, building density in the city and the types of residential structures as indicators affecting women's everyday life. The report states that decisions regarding these variables should take into consideration the needs of individual groups, including ethnic, religious and socio-economic groups. But as the report states 'there is something of a leap in moving from social needs to spatial planning variables – a gap from factors in the physical environment usually addressed by physical planners to the social needs that these physical factors must meet'. This gap between the social and the physical unfortunately characterises spatial planning in Israel. The planning system lacks 'knowledge' of social and cultural needs in planning (see Fenster 1996, 1997, 1998). The report's recommendation looks at the notion of knowledge in planning:

To avoid erroneous assumptions about the degree to which planning variables meet the needs and interests of a particular group, the group's own perception of its needs should be taken into consideration. The planner must obtain information about how various social–physical environments might lower the level of distress and improve the quality of life of the members of various population groups. Furthermore, the total impact of intervention by the planner upon decision variables in various areas must be examined. This examination must encompass the possibility of improvement in certain areas and deterioration in others, in the wake of particular policies (p. 25).

In its existing institutionalised structure the Israeli planning system does not allow other ways of 'knowledge' to be incorporated in its rational–comprehensive planning procedures, especially not the 'knowledge' of cultural and national minorities, as the examples below show. This misuse of knowledge is in fact part of the controlled nature of the planning system in Israel.

The misuse of professional knowledge as a means of control

The misuse of professional knowledge is perhaps the most explicit way of control over both Bedouin and Ethiopian groups, by means of creating planning schemes and regulations according to principles dictated by those in power (Fenster 2001, 2002). For the Bedouin living in the Negev Desert planning as control is mainly expressed in the landownership dispute between themselves and the state. As already mentioned at the beginning of the chapter the state, not recognising the Bedouin land claims, wishes to see them concentrated in a small number of towns, each with a large population, so that several tribes would live in one town.

As we have seen, this is the opposite situation to what they were used to in their traditional lifestyle, which is to live in a village-like habitation on lands they claim they own. Planning as control is expressed in the spatial differences between the modernist Bedouin town and the traditional Bedouin village. As previously mentioned, the state bureaucrats in their ambitions to protect national Jewish interests have chosen a planning solution that is inadequate and inappropriate for Bedouin cultural norms and values. The high residential density and the concentration of several tribes in one town causes a lot of discomfort for its residents, especially women.

For the Ethiopian Jewish community in Israel planning as control means they cannot choose to live where they wish, but must live in certain towns or cities. Only in these towns and cities do they get a housing mortgage from the government. Planning as control also means no real options to choose whether to live urban or rural lifestyles. This is very similar to the situation of the Bedouin, who are forced to move to urban areas.

It is clear that modernist planning in the two cases is mostly based on professional knowledge and therefore fails to meet the specific cultural needs of these communities. The misuse of professional knowledge here means that planning professionalism has been exclusive with no involvement of the community members and their local knowledge in the planning process.

It is only in the last few years that this lack of access to 'professional knowledge' has slowly started to change and the 'local based planning knowledge' has become visible as young male Bedouin begin to take the initiative and become involved in some official planning procedures. This is accomplished mainly by way of objections, for example to the official Regional Outline Plan (regarding recognition of some of their 'illegal' villages) or by establishing the 'Regional Council for the Unrecognised Villages in the Negev'. This is a local organisation that represents a very interesting combination of an official structure, 'regional council', but one related to what is perceived as illegal settlements – 'the unrecognised villages'. Most of these initiatives involve men, leaving women outside the domain of knowledge and power in decision-making processes.

The misuse of local symbolic knowledge as a means of control

Local symbolic knowledge means the local based knowledge of Bedouin and Ethiopian culture that has relevance to planning. For example, the cultural notion of tribal segregation and of women's modesty in the Bedouin case and communal lifestyle in the Ethiopian case.[5]

For local symbolic knowledge to be taken into consideration in the planning process a different planning approach must be adopted. Approaches such as the communicative model, or radical or insurgent models are more appropriate for incorporating such knowledge into the planning process. Such planning models recognise knowledge as situated and thus necessitate the involvement of community members in the planning process and in the incorporation of local symbolic knowledge in the planning process.

At the same time, the danger of misusing local symbolic knowledge seems to be high as well. Local symbolic knowledge is not free from the power relations 'game' either between the planner and the community or between the community members themselves. It obviously raises a very crucial point of the positioning of this knowledge by the planner and by the powerful people within the community. From my own experience this is a very complicated area in the planning process because, as already mentioned, knowledge is always situated, even when it comes from the community members themselves, especially if conflicts exist between them. However local symbolic knowledge, as complicated as it is, is a useful tool to understand communities' cultural construction of space, which shapes people's movements, especially women. Local symbolic knowledge is also a useful tool to use to get acquainted with local gendered knowledge and the meanings of gendered spaces.

The misuse of 'local gendered knowledge' as a means of control

Local gendered mis-knowledge means to take into account the gendered cultural construction of space. For Bedouin culture, gendered constructions of space are what are termed 'forbidden' and 'permitted' spaces. The conceptualisations of forbidden and permitted can be associated to a large extent with the Western conceptualisation of 'private' and 'public' dichotomies. In discussions with Bedouin women (see Fenster 1999a) they refer to their home and neighbourhood as 'their private place' in which they are free to move, as opposed to other areas of their town (park, town centre, etc.) that are in fact inaccessible to them. Local gendered mis-knowledge relates precisely to the fact that modernist planning does not consider such cultural constriction of space. Thus there emerges a situation of control over women in Bedouin towns because of the high residential density and the increasing likelihood of undesirable encounters for Bedouin women, affecting deeply engrained cultural codes concerning women's modesty. As a result the boundaries of forbidden and permitted spaces have become stricter in the new towns; larger areas are now designated as forbidden to Bedouin women, though they may be adjacent to permitted areas. By dictating these spaces women's access to sources of knowledge and power such as banks, libraries, shops, news agencies, health clinics, and so on is restricted. All that signifies the 'public' is actually forbidden for Bedouin women. Even walking to a public park with their children is forbidden (see Fenster 1999a). For Bedouin women knowledge is everywhere outside their homes and neighbourhoods, and by not being able to access these spaces knowledge is denied them. Thus the notion of town planning and its benefits, such as large-scale services, education, cultural activities and recreation areas, are literally not available for Bedouin women in the towns planned for them, because of local gendered misuse of knowledge by the planners. Chapter 8 suggests that these experiences of 'forbidden' and 'permitted' spaces are shared by many women living in the city, regardless of their age, personal status, class, religion, ethnicity and nationality, and lack of acknowledgement of these spaces make plans less adequate for women.

For the Ethiopian women, gendered spaces concern cultural habits of women's menstruation, which in Ethiopia dictated separation between 'pure' and 'impure' spaces. In traditional Ethiopian communities, both Christian and Jewish women's menstrual blood is considered impure, as are menstruating and post-partum women. 'Pure' and 'impure' spaces are differentiated in relation to menstruation and childbirth. This precludes menstruating and post-partum women from attending church, in the case of Christians, and from the family home in the case of Jewish women (Pankhurst 1992). This tradition removed menstruating and post-partum women from their daily routines, and during this time female relatives would bring food to the menstruation hut on special plates (see Fenster 1998). The environments into which Ethiopian immigrants to Israel moved offered no opportunities for spaces comparable with the menstruation huts, although the planners were certainly aware of traditional customs. This has caused tension among family members, with women in particular expressing feelings of guilt and shame. For them, a lack of respect for their gendered spaces is a misuse of knowledge on the part of the planners. Their needs (at least of some of them) for such separate spaces during their menstruation and after childbirth, are viewed as part of their intimate female network, a place where 'knowledge', the female knowledge, is passed. This knowledge includes daily routine, gossip, storytelling, cooking, embroidery, etc. Staying in the menstruation hut with other women creates a group, that provides both emotional and spiritual support and enables them to remain closely 'in touch' with their bodies (Doleve-Gandelman 1990). In the housing schemes in Israel such spaces were not allocated, and thus a very significant source of support, power and knowledge disappeared.

'Misuse of knowledge' actually highlights the significance of power relations in dictating which knowledge and whose knowledge is going to be part of the planning process. The complexities of the planning process as reflecting the different power relations exist in each community. The acknowledgement of such power relations raises a few ethical dilemmas concerning the incorporation of knowledge in planning.

Professional–universal vs local–relative knowledge – an ethical dilemma

We end this chapter with an ethical dilemma concerning the effects of modernist planning on cultural minorities' everyday life. This dilemma raises a very complicated and sensitive issue, especially for radical planners: whether to take into consideration cultural values even if these values abuse, for instance, women's human rights (see Fenster 1999c). It shows how the planning process can disempower community members, especially women, and how the modernist enlightenment project, which aims to equalise community members, serves the opposite purpose in the case of both Bedouin and Ethiopian women in Israel. One of the serious dilemmas that (radical or mainstream) planners should consider is how to approach cultural values in spatial planning – should planners take into consideration 'Western–universalist' or 'local–particularist' views on cultures in the planning process not only from the state's point of view but also from the point of view of some of the community members? This dilemma concerns primarily notions of control and power, of the state over minorities and of certain community members (especially men) over other members (usually women). It confronts two approaches to cultures; the first which views Western culture as superior and claims that other cultures should adjust to Western cultural norms, primarily those which assume equality of all members of the community. The second, the particularist approach, claims that different civilisations have different cultures, which need to be understood and judged within their own terms, which means acknowledging gender and class inequalities as part of those cultures.

But if planners adopt a particularist view of cultures does this mean that cultural norms which abuse women's human rights, such as the freedom of movement in space, will be taken into consideration in the planning of everyday spaces for these societies? This is of course a very problematic situation especially since we consider planning as a field in which power relations are dominant. On the one hand, modernist–universalistic plans might disempower 'communities' by being culturally blind but at the same time might empower women by assuming equality. On the other hand, particularistic plans based on local knowledge might empower communities as their cultural norms are respected but at the same time perpetuate through planning an already existing situation of women's subordination.

The two case studies of 'top-down' planning undertaken for the Bedouin and Ethiopian Jews in Israel serve as an example of such dilemmas. They show how taking a 'universalistic' view, which ignores cultural norms of a community, exacerbates the limitations of women's well-being even though these norms traditionally subordinate women. The deterioration occurred because of a lack of readiness of community members to accept changes. This dilemma reinforces the importance of using local knowledge in planning and demonstrates that planning for cultural minorities must be sensitive to such fragile situations and must consider universalist or particularistic principles in planning with the community members themselves, and between them and planners, in order to identify what is termed here 'mapping the boundaries of social change'. This means the boundaries, flexibilities, abilities, needs and wants of the different members of the community, their cultural perceptions of space and the plurality of choices they need in order to ensure a smooth process of social change.

Summary and conclusions

This chapter examined the history and role of planning traditions in the light of current flows of globalisation, critically analysing the dichotomy of professional knowledge and local knowledge combined in the planning process. The following topics were highlighted:

- **The essence of knowledge** was critically examined, distinguishing between *knowledge*, which is embodied and arises from ideas, perceptions, mental formations, information and feelings, and *ideas* that can be perceived as disembodied universal knowledge.

- **An analysis of planning traditions** with the focus on professional and local knowledge was carried out, showing on a continuum the planning traditions that are based on professional exclusive knowledge representing modernist thinking in planning and local knowledge expressed in the methodologies of alternative planning traditions.

- **The role of the planner** as promoting the incorporation of local knowledge was discussed, highlighting the importance of their ideologies and political orientation in the actual role they choose to take when the range expands between a 'technicist' mode of work to a 'community facilitator and transformator' mode of work.

- **Planning as control** has become one of the most disputed issues, especially in multi-ethnic states such as Israel. Planning as control expresses inequalities in power relations and access to influence and decision making. Three expressions of planning as control were introduced in the chapter. Each has been coined as 'misuse of knowledge': the professional, the local symbolic and the local gendered, each representing different expressions of control.

- Finally, two communities were presented, **the Bedouin** and **the Ethiopians** in Israel, both experiencing in their relationships with the state the three mechanisms of planning as control. For both communities the results of misuse of knowledge are difficult and the cause of increasing cases of poverty, in some cases increasing the level of family violation and sometimes even drug use because of the difficulties of coping with the changes.

- The last part of the chapter presented **an ethical dilemma** as to whether and how far to take into consideration cultural local knowledge that in fact abuses women's rights. The articulation of local knowledge means in the case of the Bedouin that services and parks would be allocated on a neighbourhood basis but this could also perpetuate women's lack of freedom of movement. In the case of the Ethiopians the use of local knowledge in planning means allocation of 'menstruation structures' for them but that could also mean legitimising and acknowledging the concept of women's pollution. One of the ways of dealing with such a dilemma is by involving the community members in the planning process and identifying the 'boundaries of social change' that the community members can tolerate.

The theorisation of notions of knowledge and especially local knowledge allow us to move onwards to the next chapter, in which the components of local knowledge are discussed with regard to identity issues as reproducing different ways of daily practices in the global city.

Notes

1 This is part of the Government Master Plans for the Ethiopian Absorption (1985, 1991), which allocate the mortgages to specific towns because the planners thought that it would ease the absorption of the Ethiopians if they were integrated in medium sized towns and not in large towns (see Fenster 1998).

2 I wish to thank Michel Safier for presenting his analysis on this matter.

3 This can relate to the exposure of only some of the information regarding development, so that objections are prevented, or to the misuse of information so that it
   meets the needs of several interest groups – those in power – but not of others.

4 The 20 per cent minorities in Israel consist of Muslims (76 per cent), Christians
   (15 per cent), Druzes and the other (9 per cent).

5 For elaborated analysis on the cultural norms of the two societies see also, for the
   Bedouin cultural norms: Abu Lughod (1986, 1993), Mernissi (1975, 1991),
   Lewando-Hundt (1984); for the Ethiopian cultural norms: Anteby (1999), Askenazi
   and Weingrod (1987), Doleve-Gandelman (1990), Pankhurst (1992), Salamon
   (1993), Westheimer and Kaplan (1992).

Society and space:

Diversity, difference and knowledge in the global city

CHAPTER 3

Different identities, different knowledge

In the previous chapter I discussed new ways of knowing as part of local knowledge embodied in people's everyday life and experiences in cities, a knowledge that is relevant to urban planning in the age of globalisation. In this chapter I elaborate on several aspects of such knowledge based on the assumption that space is subjectively constructed as a relative or embedded component and its embeddedness depends on people's different personal and social identities and spatial experiences.

The purpose of the chapter is threefold: to analyse how different identities create different embodied personal and social–spatial experiences in everyday life; to put these personal and social experiences and subjectivity of space in the theoretical discourse that exists in the literature; and finally to discuss the different experiences of inclusion/exclusion of people of different identities in the city.

One story that emphasises these notions of inclusion/exclusion of women from public spaces that then become forbidden and permitted boundaries relates to Arab women married to Bedouin men in the Negev, South Israel. Those women are now living in a very traditional context but in their child-hood they were raised in a more advanced environment in Arab towns where women wear Western clothes and their freedom of movement is much greater. For these women the cultural practice of wearing traditional clothing after their marriage was not easy, and on their monthly visits to their parents (who still live in Arab cities) they usually leave home wearing traditional Bedouin clothing but underneath it they wear modern clothing, a symbol of the two cultures and realities they live in. When their car gets out of the 'forbidden' spaces, that is the Bedouin town, they ask their husbands to stop the car and they change or take off their traditional dress. They mention that the boundary between the forbidden space i.e., the area where they have to dress modestly, and the permitted space i.e., the area where they can change their clothing to a modern style are usually a bit further out of the town's area where they can reappear as modernised women. Interestingly enough Muslim Bedouin women's experience in the Negev is similar to the experience of young Muslim women living in London, who 'are defined as caught between two cultures of home and school, torn by a culture clash between the secular/modern world of the school and the "traditional/fundamentalist" world of the home' (Knott and Khokher 1993 in Dwyer 1998). The latter represents the same dilemma with a much stronger 'volume' regarding the expressions of their bodies as 'contested sites of cultural representation'. This results in a process of negotiation around different styles of dress in different spaces. These examples, which are elaborated later in the chapter, emphasise the different ways gender identities are culturally constructed and how these ways affect women's experiences and knowledge and their movement in space.

Before moving to discuss how different identities effect individual symbolic construction of space let us elaborate on the connection between issues of identity and embedded, experienced knowledge. Here I use Aida Hurtado's work (1996), where she highlights some of the connections between identities and knowledge, the embodied knowledge that can not exist outside bodies (Ma Reha, 1998). First, embedded knowledge, is produced, comprehended and internalised differently for people of different identities as shown in the examples above. Certainly Bedouin and Arab women's experiences are different from those of the men in their society. In her paper Hurtado emphasises gender identities as those that have the most explicit effects on people's experiences, but she also mentions other categories such as class, race and ethnicity as significant in producing local embedded knowledge. Second, Hurtado distinguishes between personal identity and social identity. The former is part of the self and is composed of psychological traits that give us personal uniqueness. The latter consists of 'those aspects of an individual's self-concept that derive from one's knowledge of being part of categories and groups, together with the value and emotional significance attached to those memberships' (Hurtado 1996: 373). Sometimes social identity plays a stronger role then personal identity. This is probably the case with the Bedouin women, where strong cultural norms affect their lives and perhaps their personal identities as well. Third, social identity is largely socially constructed and fluid, whereas personal identity is constructed in smaller units such as families. Thus we all as human beings have personal identities, which comprise of universal processes such as loving, mating, raising children and doing productive work. Social identity is highly variable and susceptible to structural forces such as race, class, gender and sexual preference. The fourth point that Hurtado makes is that:

> the differences in value attached to significant group memberships, to a large extent determine what access individuals have to knowledge, what is considered as knowledge, and ultimately how it is that one comes to perceive oneself as knowledgeable in spite of one's group memberships. (p. 374)

Here notions of power relations and processes of stigmatisation of social identities are brought into the discussion. Bedouin women, for example, do not have much access to knowledge even at the level of higher education, or their movement in public is forbidden and thus they can not be exposed to the knowledge that exists in public spaces. This means in general that unequal values are imposed on different social identities, such as women, the poor, blacks, homosexuals, etc., and these unequal values affect their everyday life experiences and their knowledge construction.

The connection of these distinctions to our discussion is that social identities and especially their positionality and stigmatisation create the embodied knowledge, which as seen in the example above is also spatial. Thus formulating imagined spaces out of social and personal identities means constructing symbolic spaces as forbidden or permitted in the case of the Bedouin women, comfortable or uncomfortable, spaces that create a sense of belonging or not belonging, spaces of fear or security for people of different identities (Chapters 5–8). The identification of these imagined symbolic spaces is a significant part of the local knowledge needed for planning.

Before discussing the relativity and symbolism of space let us elaborate on the ways in which different identities create and effect different daily practices.

Identity and space in the city

Identity issues are becoming an important factor in global flow systems, not necessarily in parallel with the system, but mostly in reaction and contradiction to them. Borja and Castells (1997: 13) write:

> The creation and development in our societies of systems of meaning increasingly arises around identities expressed in fundamental terms. Identities that are national, territorial, regional, ethnic, religious, sex-based, and finally personal identities – the self as the irreducible identity . . . concrete fundamentalism as against abstract globalization.

As a reaction to trends of globalisation identity politics are becoming a major element in understanding experiences of exclusion and citizenship. Following the examples of the Bedouin women presented at the beginning of the chapter, two major identity constructs are first discussed: culture and gender.

Identities and the cultural construction of space

Analysing cultural construction of space necessitates acknowledging the complexities inherent in defining culture – both as an abstract and as a global concept. This creates a variety of definitions in the relevant literature, differing according to the field of research and theoretical approaches. For example, culture is defined by Cosgrove and Jackson (1987) as 'the medium through which people transform the mundane phenomenon of the material world into a world of significant symbols to which they give meaning and attach values'. Jackson (1989) defines culture as 'the level at which social groups develop distinct patterns of life'. Cultures, according to Jackson, are 'maps of meanings' through which the world is made intelligible. Overall, Mitchell (1995) presents some 20 definitions of culture, stating that in most cases 'culture is symbolic, active, constantly subject to change and riven through with relations of power'. Culture is represented in terms of spheres, maps, levels or domains. It becomes a medium of meaning and action. 'Culture is everything!'

'Culture is everything' means that cultural values influence and are influenced in almost all spaces of life, starting from the very 'private' space, the body – the self and then moving up in the hierarchy – the home, then in what is defined and culturally constructed as the 'public' space: the street, the neighbourhood and the city, and the distinction and boundary between them. Some of the analysis of the cultural constructs of these spaces appears in Parts II and III with connection to the emergence of global and local cultures (Hall 1996).

The global economy brings a culture of its own that is associated with mega-products such as Coca-Cola or McDonald's and at the same time local ethnicities and cultures emerge and allocate themselves in the making of what Hall terms 'New Cultures'. By the same token, culture could also mean how terms such as governance, community, leadership, law, democracy, patriarchy and policy are perceived as much as it could relate to norms, values of behaviour, ways of speaking, attitudes, importance of various values, etc. Cultures, argues Sibley (1995), are about implementation, about connection and relations between policy and action. It is no surprise, therefore, that these complex and contested definitions render the incorporation of cultural constructions of space difficult, especially regarding those relevant to gender relations.

Gender identity and space in the city

All historical societies have diminished the importance of women and restricted the influence of female principles. Even in Greece women's social status was restricted, as was their symbolic and practical status (Lefebvre 1992). Lefebvre's historical analysis also emphasises the male and female principles in the transformation of space. Likewise, Massey indicates male and female principles in 'spaces of modernism' (Massey 1994). These are the public spaces of the city, which are also gendered by making the distinction between public and private.

Space is where cultural values exclude women. This is because cultural construction of space has inherent in its symbolism the legitimacy to exclude women from power and influence. This section highlights these aspects of symbolism of space, which were formulated by the patriarchy. The most common are the private/public devices, which for many women in different cultural contexts such as the Bedouin mean permitted/forbidden spaces. The 'home' is the 'private' – the women's space, the space of stability, reliability and authenticity – the nostalgia for something lost that is female. 'Home is where the heart is and where the woman (mother, lover) is also' (Massey 1994). The 'public' is perceived as the white upper-middle class, heterosexual male domain. This sometimes means that women in cities in both Western and non-Western cultures simply can not wander around the streets and parks alone (Massey 1994) and in some cultures could not wander around at all (Fenster 1999a). We will return to the analysis of private/public divides in the last part of this chapter.

Symbolic spaces such as private and public are mostly relevant with regard to women, as they play major roles in the construction and defence of cultural and ethnic collectivities. They are often the symbol of a particular national collectivity, its roots and spirits (Yuval Davis 1997). Therefore women's spatial mobility is very much dictated if not controlled by these cultural symbolic meanings of space. In this way cultural and ethnic norms create 'spaces of modesty and immodesty' that then become forbidden and permitted spaces for Bedouin women in Israel as shown at the beginning of the chapter, or 'spaces of purity and impurity' for Ethiopian women as elaborated in Chapter 2. The boundaries of these spaces are usually dictated by 'the cultural guards' of society, that is, men (Fenster 1999b).

As noted above, cultural and ethnic codes might be different for different members of the community, especially men and women, because of the different social roles they play in society and because of the values of women as symbols of culture that determine collective cultural boundaries of 'us' and 'them' (Yuval Davis 1997). As analysed in Chapter 2, the 'menstruation hut' and the separation of the Ethiopian Jewish women during their monthly period or after childbirth has had the cultural function of distinguishing between spaces, of establishing an 'us–them' dichotomy that creates boundaries between modesty and immodesty, purity and impurity in Ethiopian Jewish society.

Two important issues are related to the discussion on gender, culture and space: one is the discourse around the body and the self and the other is the role of clothing as a cultural construction of space.

The 'self' and the body as cultural and gendered spaces

'The self is cultural production' argues Sibley (1995) and indeed from the very early stages of our lives our relations with the environment are culturally constructed. Object relations theory, which sets the boundaries between the 'good' and the 'bad', is also culturally constructed as norms and values of a specific culture determine them. At an early age in childhood notions of 'self' and 'other' and the boundaries between them are constructed. The discussions on the self and 'otherness' are becoming meaningful only in a specific cultural context.

How is space culturally constructed? First and foremost it is the very intimate space – our bodies – that are culturally constructed (Sibley 1995). This means looking at norms of cleanliness, dirt, odour and modes of dressing as reflecting values of shame and disgrace with body covering. Bourdieu (1984) analyses the importance of an appropriate performance of the self. He argues that 'the body is the most indisputable materialization of class taste' (Bourdieu 1984). He developed the concept 'Habitus' to describe the distinguishing aspects of behaviour, taste and consumption that are combined to create flexible rather than rigid categories of 'class'. These categories reproduce themselves by display of certain tastes and the foundations of the body, its size, volume, demeanour, ways of eating and drinking, walking, sitting, speaking, etc. It is through the bodyspace that certain cultural habits are transformed, depending on one's own identity and class. Thus through body activities the boundaries between private/public are underlined. 'Eating, like the other sin of the flesh, sex, has been constructed as a notoriously privatized activity' (Valentine 1998: 192). It belonged to the home or any other 'privatised spaces' within public domain, but not to the street. Valentine argues that the custom of eating served to regulate the boundary between the 'private' and the 'public' and what is 'forbidden' or 'permitted' in those spaces.

Looking at the human body as a cultural construct also means defining the boundaries of the 'personal space', that is, what is the distance from other persons or objects around us that we should keep so that we are not feeling threatened or intimidated. These boundaries or 'personal spaces' are culturally constructed as well. In Western societies, where values of individualism are more dominant, 'personal spaces' are wider and are stricter then in non-Western societies. See for example how close you can get to somebody else with whom you are talking, see how a person would automatically step backwards when they feel that someone else has 'intruded' into their personal space. In non-Western cultures such as the Hindu culture the range of 'personal space' is different, much more flexible, much less threatening – perhaps because norms of individualism are less dominant.

Gender, space and clothing

As the story of the Bedouin women emphasises, clothing is an important matter of identity. In the Bedouin women's story clothing became a symbol of cultural identity, symbolising the transition between tradition and modernity and vice versa, which they replaced once crossing cultural boundaries. It is very important for those women not to appear with their traditional clothing in their previous place and thus when they leave the 'forbidden space' they stop the car to change their clothes.

Clothing is a symbol of integration for immigrants too. For those who wish to integrate into their new country they have first to pay attention to their clothing: 'in novels and autobiographies', writes Jonathan Raban (1998) 'the first positive move that the immigrant makes towards assimilation is to buy himself a suit of city clothes. Before anything else, he must dress the part' (p. 46). But it is not only for immigrants that clothes play a major role in the construction of identity – it also happens in youth culture. Much of the literature emphasises the significance of fashion and styles in the creation of subcultural identities and in particular the role of dress as a contested boundary marker between different group identities (Dwyer 1998). Clothing can be perceived as the major expression of changing identities and thus it plays an important role for immigrants and young people whose identities are in transition. Clothing plays a major role for women, especially Muslim and Jewish religious women as part of their cultural constraints. We will elaborate on this last point in Chapter 8.

As seen in the example of the Bedouin women, clothing as a spatial element is expressed in Muslim women's sexuality. In Muslim culture women's bodies become contested sites of cultural representation and the role of clothing becomes significant to such an extent that 'appropriate' clothing in many cases extends the 'permission' to move in space (see Figure 3.1). Muslim sexuality is actually territorial (Mernissi 1975) and its territoriality is symbolically expressed in both women's clothing and in the articulation of 'forbidden' and 'permitted' spaces. In Iran, for example, many public areas are now segregated by sex, including schools and universities (Tohidi 1991). These restrictions are intensifying in Islamic fundamentalist countries (Beller-Hann 1995) due to the

Figure 3.1   Muslim women on London's Underground

fact that political ideals in Muslim countries use religion to restrict women's lives (Afshar 1996) rather than because the roots of the religion entail women's oppression, as commonly thought in the West.

The practice of the veil is one of the most explicit symbols of a woman's honour being intrinsic to her body. The advent of the imposition of the veil is disputed (Mernissi 1991). In several countries, such as contemporary Iran, all women, including non-Muslims and foreigners, are obliged by law to wear the veil and observe traditional Islamic hejab (complete covering of women). However, in countries where this practice is not compulsory, such as Egypt or Turkey, women also dress traditionally for various reasons, and the specific meaning attached to this practice varies according to its cultural and ideological context (Mohanty 1991). It is not possible therefore to assume, as do some Western feminists, that the mere practice of veiling women in Muslim countries indicates the universal oppression of women. In Iran, women veiled themselves during the 1979 revolution as an act of opposition to the Shah and Western culture colonisation (Tohidi 1991), while today Islamic law dictates that all Iranian women wear the veil (Afshar 1996). For many Muslim women the veil is a symbol of Islamification and of its revivalist ideals. In this case the veil is liberating, rather than an oppressive force (Afshar 1996). It allows women's access to knowledge, which usually exists in public areas. In other cases the practice of veiling solves the problem of women's desire to move freely, in 'public', while acknowledging cultural and religious restrictions (see Figure 3.2).

Figure 3.2   Muslim women in Hyde Park, London

In Cairo, Macleod (1990) observed that young, educated, middle-class women choose to dress traditionally when they leave the house to go to work, despite the fact that this code is not always demanded by the 'guards of honour' in their society (husbands, mothers- and fathers-in-law, parents). The paradoxical situation which Macleod identifies is that these women, on the edge of modernity, use the symbol of their inferiority (as judged by Western feminism) in order to broaden the space of their activities outside the home.

These examples help us to understand the function of the veil as expanding the 'boundaries of forbidden space' for Muslim women living under severe restrictions. '. . . the hejab poses the problem of understanding women's puzzling participation in the modernization process, the clash of tradition and modernity that evolved much of the contemporary world.' As such, Muslim women have succeeded in adapting to fundamentalist laws by creating laws of their own regarding the 'permitted' within the 'forbidden', by wearing the veil and the hejab. Veiled women feel less exposed to verbal or physical male abuse when using public transport or when moving anywhere in a public space (Abu Odeh 1993). Both Afshar (1996) and Abu Odeh (1993) view the veil in the context of third world feminism, as a mechanism that permits both spatial mobility for women and a sense of social and psychological safety. In the context of this chapter it can be looked at as a mechanism that allows them to access knowledge – education, employment, exposure to the media (news-papers, etc.). The veil enables women to become the observers and not the observed (Afshar 1996), while providing a short-term and immediate practical solution which, though not replacing the long-term feminist struggle for equal-ity and liberation, is not merely a sign of oppression but also allows them to retain modesty while enabling them to navigate the public arena. A culturally sensitive approach helps in the comprehension of these norms, which although viewed by Western feminists as abusive or subordinate sometimes serves as a mechanism for women to work through their practical and daily problems.

These cases are not exclusive to women in Muslim cultures. Clothing as a symbol of increasing sense of control is also known in other 'controlled' areas such as shopping malls. Youth in the Wood Green Mall in London expressed anxieties of being chased by security staff just because of the clothes they wear (Jackson 1998). Thus clothes and clothing are both mechanisms of personal and social expressions of belonging to certain cultural groups. Clothing can become a mechanism of oppression too, especially towards women and young people.

Age and the construction of symbolic spaces

In looking at two age groups old people, especially old women and teenagers, the significance of age identity in the use of space and in its symbolic construc-tion is highlighted. 'Indeed, defining life in terms of a set of "stages", write Katz and Monk, 'especially if these are linked to chronological ages, is fraught with difficulty, especially if we adopt a comparative perspective' (1993: 5). After all, it is quite obvious that youth's symbolic construction of space is totally different from old age's symbolic construction of space, if only from the obvious difference in physical ability.

It seems that the last ten years have marked a shift in the geographical research of old people, especially old women. In 1993 Katz and Monk wrote in their introduction to *Full Circles* that: 'we know relatively little about the experiences of older women around the world, even in the European Community where one-third of the female population is over fifty years of age' (p. 13). Today there is growing interest in and research on the everyday life of elderly women (Droogleever, forthcoming), with particular attention given not only to minorities or the handicapped elderly but also to healthy women who live on their own. This is mainly because life expectancy in the West is becoming higher, especially among women. Also the definition of 'old age' seems to be more elastic than ever, more and more people delay their retirement and even when retired some carry on working. There is also increasing attention paid to the function of 'golden age' in urban spaces. Planners and geographers pay attention to the special needs of this group and their interaction with the community at large. In this respect, two theories can be mentioned: 'the disengagement theory' claiming for a disengagement between old people and the community and 'the activity theory' claiming that in spite of physical and physiological changes old people have the same social and psychological needs as before, their lower level of activities are against their will and thus they should remain part of the community (Droogleever, forthcoming). Strüder (2000) suggests the heuristic model to understand why and how elderly women in Germany use space outside their home. She mentions three main elements: individual identity, income and power, and knowledge as explaining older women's activities in Germany. Churchman (1993) identifies the needs of old people in urban spaces as similar to those of women, children and physically challenged people. She says that the larger the residential area and the more variety of services, housing and employment opportunities it can offer the better it is for these groups. There is thus an advantage to residential areas where services are located within walking distance and where public transport is accessible, easy and frequent.

The literature around young people is much more popular, although until recently geographers have been criticised for ignoring children and excluding the experience of youth culture from their research (Valentine *et al.* 1998). David Sibley's book *Geographies of Exclusion* (1995) deals with youth culture and space and the sometimes vague boundaries between the 'permitted' and the 'forbidden'. He makes a clear connection between the social identity of youth and their spatial movement, looking at border crossing and zones of ambiguity in spatial and social categorisations: 'There is always some uncertainty about where the edge of Category A turns into the edge of Category not-A' (Leach 1976). Child/adult categorisation illustrates such a contested boundary. Who is a 'child' or a 'teenager' varies in different cultures, and the age range of these categories has changed considerably through history within Western, capitalist societies. Thus the boundary separating child and adult is a fuzzy one and perhaps therefore the most ambiguous period in life is perceived as adolescence. Adolescents are not yet adults but they are not children anymore. They are denied access to the adult world but they attempt to distance themselves from the world of the child. Adolescents may be threatening to adults because they transgress the adult/child boundary and appear discrepant in the 'adult' spaces. Sibley argues that 'the problems encountered by teenagers demonstrate that the act of drawing the line in the construction of discrete categories interrupts what is naturally continuous' (1995: 35). This is part of the question Valentine, Skelton and Chambers (1998) discuss in their introduction to *Cool Places*. What does it mean to be young, they ask, and discuss several aspects of the meaning of youth: work, having a good time, being rebellious, respect for adults, innocence, etc. The case studies in their book explore the diversity in young people's lives, their place on the geographical map and their relevance to geographical debates. Like Sibley, they argue that it is not easy to define youth: 'adolescence has been invented to create a breathing space between the golden age of "innocent" childhood and the realities of adulthood' (p. 4) and like Sibley they focus on the ambiguous nature of adolescence. Because of the cultural constraints on adolescents the street becomes the only autonomous space that young people can carve out for themselves and hanging around, and larking about, on the street, in parks and in shopping malls is one form of youth resistance to adult power' (Valentine *et al.* 1998: 7).

Physical and mental challenge in the city

Physical and mental challenge is a PR term for 'disability', 'crippled people', 'the crazy', 'lunatics' and other expressions, which were invented through history to exclude and oppress disabled people. Gleeson (1998) argues that it is impossible to define disability objectively because the term has many different uses in various places. But what is universal is the oppression of disabled people together with other 'marginal identities' such as women, gays and indigenous people (Young 1990). Gleeson (1998) identifies three principal aspects of the urban social oppression of disabled people: inaccessibility, poverty and socio-spatial exclusion. Disabled people develop symbolic spaces as well as do other identities. Because of a lack of accessibility to certain spaces those spaces become 'uncomfortable', 'forbidden' or 'dangerous'. Batler (1998) mentions two problematic issues that complement Gleeson's categorisation of oppression: lack of understanding from the 'others', in this case the physically healthy people about the problems of spatial movement for disabled people, and the outright discrimination and inadequate resources and services that disabled people of all ages must face in their social, political and economic lives because of this lack of understanding. This social exclusion and marginality becomes spatial exclusion and marginality (Sibley 1995).

An example of the combined situation of disabled people is the location of the two clubs for blind people in Tel Aviv and Jerusalem. These clubs are located in marginalised areas with difficult accessibility, which is in fact the opposite of what the blind need in order to feel comfortable and safe. Blind people mentioned in interviews[1] that what makes spaces comfortable for them is their familiarity with them and their easy accessibility. The more comfortable they feel the more they use public spaces. For some of them the differences between private and public spaces are reflected in terms of their control and familiarity with space – in what they term private space they can control the order of things and move easily, whereas in public spaces they can not. They also mentioned their desire for spatial inclusion as a major component of their integration in society. Some of them emphasised their psychological need to feel and know that they contribute to society at large by being able to take part in public activities. However, because of their perception of public spaces as full of 'many obstacles' they do not go out a lot.[2] Physically disabled people described their feelings of oppression similarly to Gleeson's (1998) findings: most of them do not work, they feel excluded and thus they do not often go out of their home because of the lack of physical accessibility to many public places such as restaurants, cinemas and theatres: 'This means that the physical layout of cities – including both macro land use patterns and the internal design of buildings – discriminates against disabled people by not accounting for their mobility requirements' (Gleeson, 1998: 91).

The points Gleeson specifies as to what physical inaccessibility means were mentioned both by blind and physically disabled people living in Israel: physical barriers to movement for disabled people, an architecture style that prevents entry to anyone unable to use stairs, and public transport modes. Such discriminatory design, argues Gleeson, is the primary reason for exclusion and oppression of disabled people in all aspects of life: exclusion from employment, exclusion from participation in political life, exclusion from everyday life in the city.

Sexual preferences and the construction of symbolic spaces

'Like any other oppressed groups', writes Leonie Sandercock in the introduction to her *Making the Invisible Visible* (1998b: 12), 'gays and lesbians have stories to tell about the ways in which their lives in cities and neighborhoods have been and are impinged on by social and spatial policies and how they, in turn, respond to and contest certain policies'. Gays and lesbians, argues Sandercock, have particular needs with respect to urban services and spatial policies and one of the questions she poses is whether planners can help to create safer streets and neighbourhoods for gays and lesbians and how their life experience can become part of planning procedures. Duncan (1996) discusses norms around 'forbidden' and 'permitted' behaviour of homosexuals. She notes that surveys have shown that the majority of respondents have no objection to homosexuals as long as they do not express their sexuality in pub-lic. Valentine (1993) makes this distinction between forbidden–permitted/public–private and notes that suburban housing developments as sites of overtly heterosexual as well as familial sentiments and rituals are generally considered alienating environments by lesbians and gays and in such areas even the home can become an alienating site.

Does this lead to the ghettoisation of gays and lesbians as already happens in San Francisco, Paris, London and other places? Duncan (1996) advocates the opposite. She suggests that:

> lesbian and gay practices which potentially denaturalize the sexuality of public places could be more effective if they were widely publicized. If they were made more explicit and readable then contests around sexuality would become more visible to the straight population'. (p. 138)

How to do that? By practising 'deconstructive spatial tactics', which can take forms of marches, Gay Pride parades, public protests, art performance and street theatre as well as overtly homosexual behaviour such as kissing in the street (Duncan 1996).

In this section we have discussed people's personal experiences deriving from their different identities and their symbolic construction of space as a result of these experiences. I move now to analyse how symbolism/relativity of space is constructed for different personal and social identities and to elaborate on the dialectics that exist in the literature between absolute and relative space.

The dialectics between absolute and relative space

Analysing the symbolic meanings of space as perceived by people of diversity living in global urban spaces is important in understanding how to manage and plan these cities, especially in the era of changing global economies, which put businesses in the forefront of cities' activities and tend to undermine people's perceptions, values and needs. As I elaborated in Chapter 2, current planning traditions, both mainstream and alternative, do not take into consideration this kind of local knowledge as part and parcel of identifying people's needs in the planning process. On the contrary, planning in the global era assumes spaces to be only physical and 'absolute'.

Notions of the relativity and symbolism of space as embedded in our every-day knowledge are discussed in what was labelled by Healey (1997) as: 'the relational understanding of social processes'. This approach acknowledges the relativity of our meanings of reality: 'they are interpretations based on pre-existing knowledge, our cultural predispositions as well as observation and experimentation' (Healey 1997: 119). Massey (1994: 2) also argues that space is not an independent dimension but is constructed out of social relations: 'the spatial is social relations stretched out'. Social relations of space are experienced differently, creating a simultaneous multiplicity of spaces for people of different identities who have different experiences of spatial movement, abilities and perceptions. The spatial for Massey:

> 'can be seen as constructed out of the multiplicity of social relations across all spatial scales, from the global reach of finance and telecommunications, through the geography of tentacles of national political power, to the social relations within the town, the settlement, the household and the workplace'. (1997: 4)

Let us now elaborate on the dialectics that exist in the literature between absolute and relative space and then move on to discuss how the relative meanings of space are symbolically constructed for people of different identities as forms of exclusion and inclusion in globalised urban spaces.

From the absolute to the relative thinking on space

The role of space in social theory has become increasingly significant in the past three decades. Previously the two disciplines, geography and sociology, worked separately (Madanipour 1996). Thus geographers analysed spatial structures solely from the geographical perspective, despite their clear social impact on one another (location theory and the von Thunen's model of agricultural land use are two such examples). At the same time sociological works such as Alfred Weber's industrial location theory and the 'ring mosaic of urban land use', identified by the Chicago School, did not incorporate spatial analysis at all.

The linkage between space and social relations finally emerged as a result of a series of developments, for example the emergence of new spatial structures of combined and uneven development, the changing structures of social relations, and the formation of class and non-class social movements (Gregory and Urry 1985). Henceforth it has become clear that spatial relations can represent and reproduce social relations. One expression of this shift is the silent but clear acknowledgment by the Marxists that geographical, and especially spatial perspectives are vital for understanding social theory. Crises of capitalism not only have their own geographies but also recognise that spatial structures are intrinsic to the resolutions of these crises (Harvey 1989). The process of 'the spatialization of social theory' (Soja 1989) began to attract attention precisely as Marxism itself became subject to increasingly critical analysis by geographers. This was further reinforced by Giddens (1979, 1981), whose structurationist theory disentangled the Marxist dialectic, separating 'structure' and 'human agency' in order to investigate their interrelationship (Smith 1994). Another important change, which influenced the introduction of space in social theory, was the shift in attitudes towards modernity. Modernity, the major way of thinking during this era, began to be dismantled in the 1960s. As shown in chapter 2 modernity, in turn, initiated significant shifts in the nature and interpretation of development, planning and economic growth. Some argue that planning and economic growth became interpreted as exploitation of marginalised groups, such as blacks or other ethnic minorities, women and students, all of them expressing discontent with their subordinate position within the modern order (Lefebvre 1992). This spatialisation of the Marxist approach, a criticism of modernity, is echoed in planning traditions too, as explained in Chapter 2. There I mentioned that the process of planning may be viewed as one of the areas in which power relations are expressed, in that the dominant group controls the resources – material resources as well as symbolic resources – knowledge. The incorporation of space into social theory has sharpened the dialectics and the discourse existing in the literature around the relativity and symbolism of space as reflections of social identity and relations.

Absolute and relative spaces – notions of representation and symbolism

Because of the significance of this subject as being part of local knowledge in planning I shall elaborate on the ways philosophy and physics deal with it before moving on to discuss its geographical perspective.

The philosophical debates about space have been dominated by a dichotomy between absolute versus relational theories. Madanipour (1996) elaborates on the theory of absolute space developed by Newton, who saw space and time as real things, as: 'containers of infinite extension or duration. Within them, the whole succession of natural events in the world find a definite position' (Madanipour, 1996: 5). Alongside that, the relationist theories were developed as a critique of the concept of absolute space with the first major opposition made by Leibniz: 'who held that space merely consisted in relations between non-spatial, mental items' (Madanipour, 1996: 5). Kant saw spaces as:

'belonging to the subject constitution of the mind and not an empirical concept derived from outward experiences. 'Beyond our subjective condition, the representation of space has no meaning whatsoever . . . space and time cannot exist in themselves but only in us . . . so outward objects are just representations of our sensibility whose form is space'. (Madanipour 1996: 5)

The contribution of physics to this debate is through Einstein's opposition to the distinction between absolute and relative space. Instead he perceives space as both 'positional quality of the world of material objects' and 'container of all material objects', where the former is rooted in the concept of place, the material place, and the latter is a more abstract meaning, seeing space as unlimited in extent. Colquhoun (1989 in Madanipour 1996) distinguishes between social space defined as the spatial implications on social institutions and urban space defined by Krier (1979 in Madanipour 1996) as 'external space' that is, all types of space between buildings in towns and other localities.

Another distinction is between the real and the mental space (Madanipour 1996). Real space as understood through the senses is differentiated from human beings' intellectual interpretations of the world, which create a mental construct. This is the epistemological basis to the use of people's senses and perception of space – their mental cognition and mental maps as a methodology for identifying local knowledge (elaborated on in Part III).

The dialectics between absolute and relative space exist in geography as well. Relative space is used for what philosophy calls relational space and the absolute space refers to the physical and eminently real or empirical space. Absolute space has been defined by several researchers as an objective and real space while relative space has been defined as perceptual and socially produced (Madanipour 1996). The re-emerging of the discourse around space was started during the 1970s by postmodern theorists who showed interest in the dilemma between physical space: 'which can be understood immediately by the senses, and mental space, which needs to be interpreted intellectually' (Madanipour 1996: 9). But, asks Madanipour, 'what do we think of this dilemma between the mass and the void in dealing with urban space?' (p. 10). Does this dichotomy have any meaning in urban everyday life? Or as Madanipour put it:

> Does it make sense to say that in our walking in the street we have both a spatial experience, in which enclosures are different from open spaces and streets are different from squares, and an experience of the material objects which shape or condition this space? (p. 10)

For Raban it does make a difference. 'For me' writes Raban (1998: 162), 'this [the city symbolism] gives city life a curiously vertiginous feeling. I move on the streets always a little apprehensive that the whole slender crust of symbolic meaning might give way under my feet'. In walking in the streets, shopping in the city centres or enjoying ourselves in parks, although we use, see and sense the *same* built, physical, absolute space each of us perceives it, uses it, views it *differently*, and this difference is a result of our personal and social identities that make these perceptions of space similar for certain identities (youth, the elderly, women) but also different (black versus white in some parts of London or Palestinian versus Jews in some parts of Jerusalem). For me this is the most important distinction between the absolute and the relative space and this is how I interpret these two elements. The personal and social constructions of space of the different groups are in fact the 'local embodied knowledge', which is analysed in Part II.

One can not finish the discussion on absolute and relative space without mentioning the writings of Henri Lefebvre (1992), the great French philosopher and activist who interlinks in his thought the two themes of urbanisation and the production of space, although denying the city any kind of meaningful entity in modern life. Lefebvre focuses on the process of urbanisation or the production of space that binds together global and local, the city and the country, the centre and the periphery in new, unfamiliar and dialectical ways (Harvey 1996). Lefebvre states that even anthropology acknowledges how space is a reflection of society:

> the space occupied by any particular 'primitive' group corresponds to the hierarchical classification of the group's members, and how it serves to render that order always actual, always present. The members of archaic societies obey social norms without knowing it – that is to say, without recognizing those norms as such. Rather they live them spatially. (Lefebvre 1992: 229–30)

This is perhaps one of the best ways to emphasise the universal connection between social order and space. Spaces, argues Lefebvre, differ in their degree of 'participation' in nature, that is, between the cosmic naturalness (air, water, sun, 'green space') and the genitality (the family unit and biological reproduction). These two elements help make a distinction between two ways of approaching space: *representation of space*, that is, 'the cosmic' as represented in maps and plans, transport and communication systems, information conveyed by images and signs, and *representational space*, that is, the genitality, which for Lefebvre is nature and fertility: 'as directly lived through its associated images and symbols, and hence the space of inhabitants and users, a space understood through non-verbal means' (p. 39). A space, then, that perceives and contains local knowledge, the knowledge we look for in order to plan our cities so that they become better places to live.

Symbols and symbolisms are, according to Lefebvre, much discussed but not intelligently, as some of the symbols had a material and concrete existence before coming to symbolise anything. For example, the labyrinth was originally a military and political structure designed to trap enemies in a maze. So, initially and fundamentally absolute space has a relative aspect and relative spaces for their part encompass the absolute. They exist in dialectics that the one actually is part of the other. The same dialectical relationships exist between the city and the country and between urban and agrarian spaces. The city or the town has a two-sided relationship to the country: it consumes the surplus products of the rural areas and at the same time provides administrative and political protection. Lefebvre sees these symbiotic relationships as maternal (the town stores, stocks) and masculine (the city protects while exploiting or exploits while protecting) – the town and its surroundings thus constitute a texture, says Lefebvre, in which absolute space assumes meanings addressed not to the intellect but to the body, meanings conveyed by threats, by sanctions, by a continual putting to the test of the emotions. It is a representational space rather than a representation of space.

Philosophically, absolute space is located nowhere. It has no place because it embodies all places, and has a strictly symbolic existence. Absolute space is a space of sanctuary: a temple, church, tombs. In ancient Greek absolute space may contain nothing but in the West absolute space has assumed a strict form, that of volume carefully measured, empty, hermetic and constitutive of the rational unity of Logos and Cosmos. It embodies a coherent stability. The Romans, on the contrary, reintroduced difference, relativity, and varying (and hence civil) aims into a Greek space. For the Romans:

> The city constituted their representation of space as a whole, of the earth, of the world. Within the city, on the other hand, representational spaces would develop: women, servants, slaves, children – all had their own times, their own spaces. The free citizen – or political soldier envisioned the order of the world as spatially embodied and portrayed in his city. (Lefebvre 1992: 244)

Rome offers the production of 'space of power' and not only in its absolute terms, but also in its images, symbols, and the constitution of buildings, of towns, and of localised social relationships. As a Marxist, Lefebvre illustrates the dialectic relationships between absolute and imaginary spaces. He uses the symbiotic relationships between the city and the country as an example to emphasize the fact that each can be perceived as absolute and imaginary and the dialectics between them mean that the city cannot exist without the country and vice versa. Absolute space is imaginary but also 'imaginary' spaces such as religious spaces have a 'real' political existence. In this respect he in fact integrates mental space into its social and physical context, arguing that all of these dimensions should not be kept separate, but be combined into a 'unitary theory of space'. Social space is a social product so that every society produces its own space and it is this production of space that should be of interest rather than the objects and things in space. Such production of space is the basis for the layout of towns and regions and for the organisation of the environment. Here is probably the right place to mention the distinction made by Soja (1989) between physical space of material nature and the mental space of cognition and representation. The latter includes attempts to explore the personal meaning and symbolic contents of mental maps and landscape imagery. These meanings affect and to a large extent dictate people's movement and use of the city's spaces. This is what Raban (1998) calls: 'the private city', the personal, intimate city we all construct out of our daily experiences.

In fact the dialectics between absolute and relative space are also the dialectics between the production of space as a social product and the creation of a social space, which reflects social relations and power in society. But in fact these dialectics are not between two but between three elements – a triad: 'the dialectical relationships which exists within the triad of the perceived, conceived and the lived' (Lefebvre 1992: 39, 40) (in spatial terms: spatial practice, representations of space, representational spaces).[3] *The perceived* is what Lefebvre terms *spatial practice*, which means the dialectics between 'daily reality (daily routine) and urban reality (the routes and networks that link the places set aside for work, "private" life and leisure)' (p. 38). Spatial practice means our daily spatial experiences, which take place in already, planned spaces. In neocapitalist society it is the relations between the private daily spatial experience and the urban reality of social and physical networks that function in urban spaces. *The conceived* is what Lefebvre terms *representation of space*. This is 'the space of the scientists, planners, urbanists, technocratic subdivides and social engineers, all of whom identify what is lived, what is perceived and what is conceived' (p. 38). This is the spatial product of the *professional planning knowledge*, the space that is conceptualised, divided and networked using the professional knowledge of the planners, architects, urbanists, etc. As Lefebvre mentions it is different from *the lived* but it has an impact on the lived and the perceived.

The lived is: 'space as directly *lived* through its associated images and symbols, and hence the space of "inhabitants" and "users" . . .'. This is the dominated and hence passively experienced space. It overlays physical space, making symbolic use of its objects' (Lefebvre, 1992: 39). This is actually the mental space, *representational spaces* that can be expressed in non-verbal symbols and signs. The product of this *lived space* or *lived experience* is what I term in this book *local embodied knowledge*, expressed in people's narratives and mental maps (see Parts II and III).

Taking Lefebvre's triad model, I would argue that the book focuses on the dialectics between the *conceived* and the *lived* experiences or *the representation of space* and *representational spaces*. In my own terminology these are the dialectics between *the professional* and *the local embodied knowledge*. These dialectics express themselves in the third angle: *the perceived space* or *the spatial practice*. Thus the formulation, design, planning and management of *the conceived* can not be carried out without understanding *the lived – the representational spaces*. Parts II, III and IV present some of this body of knowledge.

In the remaining section of this chapter I discuss how people's social experiences, different identities and different constructions of space affect the way they are excluded or included in the city.

Exclusion/inclusion and forms of citizenship in the global city – notions of culture, gender and power

Exclusion/inclusion in the global city

The discourse around social and spatial inclusion/exclusion in the global city is connected to the dialectics around the relativity and the absolute perception of space. Spaces are physical or absolute for everybody yet they become relative or symbolic, that is, forbidden or permitted for some and not for others. Exclusion of whom? Of the 'other', of women, the working class, homeless, blacks, non-natives or the mob (McDowell 1999). Exclusion is usually spatial, it sets boundaries in spaces where the 'excluded' cannot enter, these boundaries are sometimes material and sometimes symbolic. It is usually state agencies and antagonistic communities in the dominant societies that have the power and the capacity to affect the lives of the minorities by excluding them from public spaces or socio-economic or political activities. Social exclusions are usually spatial so that the image of the community becomes the image of a place (Sibley 1995). The power to exclude spatially is actually the power of planning, of monopolising space through zoning for example, and the relegation of weaker groups in society to less desirable and attractive spaces. One such example is the nineteenth-century arcades in London, privately owned spaces in terms of the regulations of spatial behaviour, which were dictated by their owners. These regulations in fact increase control and constraints on public behaviour in those privately owned spaces. Because of their exclusionary tone in the nineteenth century they were considered safe places for upper-class women (Rendell 1998). Even today these arcades can be perceived as semi-private in that there are sets of rules the users have to follow. In the Burlington Arcade in London these rules appear at the entrance. You are expected to obey its Regency laws, still imposed by top-hatted beadles, forbidding you to whistle, sing or hurry. These rules have sustained a high standard of behaviour since Londoners first went shopping in the Burlington Arcade and distinguish bodily behaviour between the street and the arcade, making it a semi-exclusionary place.

As mentioned in Chapter 2, those who control space via planning are those who have the power to exclude. In this way spatial structures can weaken or strengthen social boundaries of inclusion and exclusion. The design of cities had in various periods an instrumental role in the exercise of power.

Power in planning is expressed in grand design as much as in simple geometry (Sibley 1995). In this respect one can observe a strong connection between a strong classification of space and the rejection of social groups who are non-conforming. Being non-conforming they become a threat to social order and as such must be controlled and feverishly excluded. Women in certain cultures are excluded for the same reasons in order to keep their honour and dignity, thus creating the private/public dichotomy, which then symbolises the forbidden/permitted dichotomy (Fenster 1998, 1999b).

The more exclusionary public spaces become the more classified and controlled they are. Sibley (1995) argues that in order to understand problems of exclusion in modern societies we need: 'a cultural reading of space or anthropology of space'. Such a cultural reading of space emphasises the rituals of spatial organisation of the 'sacred'. These rituals are part of 'a series of constantly contested and negotiated social practices whose meanings are influenced by the power and status of their interpreters and participants' (Rao 1995: 173). A cultural reading of space means identifying the exercise of 'boundary-drawing' and its roots (Massey 1994). In global cities these forms of exclusion become more explicit in terms of deprivation and diversity. The globalisation of the economy and its resultant acceleration of urbanism have increased the ethnic and cultural diversity of cities through national and international migration forces and the intensification of exclusion in urban spaces

(Borja and Castells 1997). These forms of exclusion are both spatial and legal. The spatial expressions of such exclusions happen because this low-skilled and low-paid working class, which serves the international service sectors, exist and live alongside but well separated from the community of global professionals (Clark, 1996). Global cities then become places of exceptional wealth and affluence as much as places of severe disadvantage, deprivation and exclusion. From legal perspectives forms of exclusion are expressed mainly in the various definitions of citizenship and residency in global cities and states, as shown in the next section. These processes are expressed in a growing urban ethnic segregation in all global cities. In London, for example, 42 per cent of the population are ethnic minorities, concentrated in certain districts that are characterised by a lower level of education, higher unemployment rate and lower economic activity (Borja and Castells 1997). In 1997 over 110,000 households in London were homeless, and there was an estimated shortfall of over half a million affordable homes for rent. As a result Prime Minister Tony Blair announced the establishment of a special unit to tackle problems of 'social exclusion in Britain' as this was announced by Peter Mandelson, the Minister without Portfolio, as the 'greatest social crisis in our times' (Kettle and Moran 1999). In what follows the legal–citizenship aspects and the spatial–private/public aspects of exclusion will be elaborated on.

Citizenship in global urban spaces

Legitimised forms of exclusion are expressed in the different definitions of citizenship. These definitions are identity related in that they dictate which are included within the hegemonic community and which are excluded. These definitions could have negative effects on women, children, immigrants, people of ethnic and racial minorities, gays and lesbians and sometimes on elderly people too. Citizenship definitions are also spatial. They dictate in which *representations of space* the rights and duties of a citizen are relevant and in which spaces they are not. One such spatial distinction is the separation between private and public spaces, as shown later. As global urban spaces become more and more diversified in terms of their citizen's identities these notions of citizenship are becoming more crucial and relevant to the discussion on the future of global cities.

The various definitions of citizenship can be viewed as one of the legitimate ways to exclude 'strangers' by way of clarifying the boundaries between 'us' and 'them'. A conflict is inherent here, the recognition that 'strangers' are in fact a socially diverse group of people with different abilities and needs, and the strong belief that in a democratic society we should all have equal access to all goods and resources that society offers and that we should develop the notion of 'the citizen of the global city'. Many critics from both left and right recognise that citizenship by definition is about exclusion rather than inclusion for many people despite the common definitions of the term (McDowell,

1999). Let us have a look at the current work on citizenship both as inclusion-ary and exclusionary with the purpose of drawing out implications on city life and perhaps identifying new forms of citizenship in the global city.

Popular definitions of citizenship mention equality, communality and homogeneity as part of what citizenship means, almost in contrast to notions of difference and cultural, ethnic and gender or racial diversity. Citizenship is interpreted by Marshall (1950, 1975, 1981) as 'full membership in a com-munity', encompassing civil, political and social rights. The discussion on citizenship during the past decade is viewed by many as the result of political and social crises, wherein the exercise of power is challenged and thus the widely used definition of citizenship has shifted to a more complex, sophist-icated, less optimistic interpretation of exclusion (Kofman 1995). The idea of citizenship is now used analytically, to expose differences in the *de jure* and de facto rights of different groups within and between nation states (Smith 1994). The concept is also used normatively to determine how a society, sensitive to human rights, should appear.

Expressions of citizenship in space have been coined as 'spaces of citizen-ship' (Painter and Philo 1995), referring to the expression in space of the rela-tionship between the state and its citizens. It means approaching citizenship from its social and political aspects of rights and participation and defining spaces of inclusion and exclusion. Another interpretation of spaces of citizen-ship is the quest for equality, meaning whether all citizens get equal treatment from the state in matters which involve equal access to resources, that is, the provision of equal access to natural resources such as land, water or minerals, as well as equal access to infrastructure, welfare services, education, employ-ment and knowledge. This relates both to cases of discrimination of those who are defined legally as citizens but do not receive equal level of services, such as the Bedouin and the Ethiopians in Israel, and those whose citizenship is denied.

City life in between equality and difference

The notion of difference in community life in globalised urban spaces is expressed in Iris Marion Young's work. She develops what she calls 'an ideal of city life as a vision of social relations affirming group difference' (1990: 227). She argues:

> If city politics is to be democratic and not dominated by the point of view of one group, it must be a politics that takes account of and provides [a] voice for the dif-ferent groups that dwell together in the city without forming a community.

However, in reality, cities and their citizens are powerless before the domina-tion of 'corporate capital and state bureaucracy'. Moreover, privatised decision making in cities reproduces and exacerbates inequalities and oppressions, segregations and exclusions. Young proposes to construct a normative ideal of city life as an alternative to both the ideal of community and the liberal individualism she criticises as a-social. Ideal city life for her is 'being together as strangers' (p. 237), where groups do not stand in terms of exclusion and inclusion, but 'overlap and intermingle without becoming homogeneous' (p. 237) and where 'everyone can speak and anyone can listen' (p. 240) in public spaces. Social justice in such an ideal city requires the realisation of politics of difference, which recognise and affirm diverse social groups by giving political representation and celebrating their distinctive characters and cultures. In her conclusion she argues that: 'social justice involving equality among groups who recognize and affirm one another in their specificity can best be realized in our society through large regional governments with mechanisms for representing immediate neighborhoods and towns' (p. 248). This option opposes the idea of decentralisation of urban decision making and the creation of small autonomous local communities suggested as a way to manage city life.

It is interesting to compare Young's model of large regional government to what Innes (1998) terms 'a mature democracy' that can tolerate the aspirations of minority groups and provide conditions for such groups to flourish by sustaining their identity. Within this democratic perspective Innes argues that separateness can be a sign of 'inclusion'.

Citizenship in the global city – notions of comfort, belonging and commitment: towards a new definition

From the above basic statements different view of citizenship in globalised urban spaces can be deduced. Yuval Davis (2003) suggests the notion of 'multi-layered citizenship', which means that one's citizenship in collectivities can be identified in different layers – local, ethnic, national, state, cross- and supra-state. One's citizenship is affected and often constructed by the relationships and positionings of each layer to the others in specific historical contexts.

Moving onwards in the discussion another perspective of 'citizenship in the global city', is suggested here. This definition emphasises the closely related connections between citizen rights and urban planning. The three concepts of comfort, belonging and commitment are interpreted in what follows in the book into citizen rights: 'the right to feel comfort in the city', 'the right to belong to the city' and 'the right to feel committed to the city'. These three sets of rights represent the essence of a good quality of life in the city. The definition and analysis of each set of rights follows in Chapters 5–7 and comes out of people's own experiences in each of the cities. What we can mention here is the fact that the fulfilment of each set of rights is connected to planning and management of our cities. In Chapters 5–7 we elaborate on these notions especially as they relate to the following questions: how do forms of belonging that are different than those of the majority hegemonic become part of 'the right to belong'? The example of the Palestinians in Jerusalem or the Bangladeshi in London emerges in this discussion. Who defines these spaces? And ultimately, what creates commitment in citizens to their cities? Probably part of this commitment is created when their citizen rights for comfort and belonging are fulfilled. These are some of the basic premises that Parts II and III explore in looking at the narratives of the citizens of Jerusalem and London regarding comfort, belonging and commitment.

While 'the right to comfort, belonging and commitment' will be articulated later we can still conceptualise the notion of 'citizenship in the global city' in the context of the three elements looking at the already discussed dichotomy of the right to be equal with the right to maintain difference. Elsewhere (Fenster 1996) I defined 'pluralist citizenship' as providing *similar* treatment in *similar* situations and *different* treatment in *different* situations. That is, citizenship is inclusive to all members of the community whatever their identities. In this way principles of equality are met in terms of equal access to services or other resources that the state provides to its citizens. But at the same time 'pluralist citizenship' is also exclusive. It enables the acknowledgement of difference, especially on an ethnic and national or any other identity base, thus providing *different* treatment where *different* identities are at focus, thereby protecting the principles of uniqueness, such as preserving customs of inheritance and landownership, providing budgetary support for religious needs, acknowledging cultural norms and values of ethnic minorities[4] (see Fenster 1996, 1997, 1998), protecting the rights of homosexuals, providing services for disabled people and promoting gender equality.

Let us now connect between the two terms. Understanding the meanings of 'citizenship in a global city' in the context of the right to comfort in the city, the right to belong to the city and the right to feel committed to the city provides a wider perspective to the notions of equality and difference. The right to maintain and respect difference is connected to the right to belong, if one's difference is recognised and respected one's sense of belonging is increased, or if cultural norms and values of communities are respected their members feel more committed. But a wider definition of 'citizenship in the global city' would take these concepts further in the discussion on the field of urban planning to what Sandercock (1998a) calls 'expanding the language of planning'. 'Citizenship in a global city' thus includes the duties of states and cities to their citizens to design 'the city of memory, of desire, of spirit as opposed to the planners' rational city' (Sandercock 1998a: 207). To design such cities planners have to understand what makes urban spaces more comfortable, what increases a sense of belonging and a sense of commitment. Parts II and III suggest new ways of understanding these concepts.

The private/public discourse

The private/public distinction marks the boundaries of visible and invisible citizen rights. Citizen rights are blurred by what are termed 'private' spaces, especially 'at home'. Those who are most affected by these distinctions are usually women and children, whose rights are mostly abused at the home level. The meanings that divide between 'private' and 'public' are products of the hegemony and the patriarchy (Jackson 1989), which for certain groups in a community, especially women, children and people of ethnic minorities, dictate the boundaries between the 'forbidden' and the 'permitted' spaces and have enormous effects on their citizen rights. The cultural distinction between the private and the public is one of the examples of *representational spaces* when absolute spaces become relative and symbolic.

Some of the most brutal and cruel cases of human and citizen rights abuse are connected to these definitions of private/public as forbidden/permitted. These are mainly expressed in the lack of freedom to move in space and/or imprisonment at home – whether enforced physically or psychologically through fear and terrorism or imposed by rules and cultural meanings of spaces. All prevent women's freedom of movement in space (Fenster 1999b).

The private and public divide derives from Western liberal thought and has its roots in the transformation of the hegemonic power relations in the society from a patriarchy, in which the father rules over both other men and women, to the fraternity, in which men get the right to rule over their women in the private domestic sphere, but agree on a contract of a social order of equality among themselves within the public, political sphere (Pateman 1988, 1989).

Some feminists define 'private' spaces within the human and citizen rights discourse as first and foremost the right of women in relation to their private 'territories' – their bodies. This includes the right to prevent violence against their bodies such as rape, sexual harassment, forced prostitution, the fight against discriminatory family and marriage laws and against any restrictions on women's reproductive rights and the condemnation of any persecution on the basis of sexual orientation. To fight these abuses feminist activists are arguing that the dichotomy of public/private is false, invoked largely to justify female subordination and to exclude abuses of human and citizen rights in the home from public discourse. They further argue that what occurs in private spheres shapes women's abilities to participate fully in public sphere activities as citizens and that women who are subordinated in private spheres cannot function as full and equal citizens in public matters (Bunch 1995).

The changing meanings and symbols of private/public spaces in the global city

New practices and new scales of organisation within global capitalism are currently redefining the public/private divide in both first and third world countries. Mega-corporations are replacing the power of governments as economies are built on a global scale. In this context private/public divides are no longer clear-cut (Eisenstein 1996) because services, spaces and activities such as health, child and elderly care that used to be 'public' in industrialised welfare states are privatised. In developing countries the opposite is occurring:

what was the duty of the family is now in the hands of the state (child and elderly care in third world countries) while global corporations have taken over employment that was formerly locally organised. The processes of transnational economic intrusion into third world countries means in many cases further exploitation of women, as poverty-wage workers, particularly in factories. Another example of the 'invasion' of the 'public' into the 'private' is the growing practice of home working, which further exploits women's labour.

The notion of 'public' (as well as 'private') is not only a gendered matter but also a race and class matter (Sullivan 1995). In many cases the notion of 'public' interest is being replaced by the notion of 'the wealthier upper class public interest' (Eisenstein 1996). Public spaces are renegotiated in the interests of only some of the 'public' and power relations define the boundaries between private and public. A clear example of this is the disinvestment in public services such as playgrounds or parks in poor and minority neighbourhoods while tax breaks and other mechanisms facilitate gentrification and up-scale developments for the affluent 'public'. Another example is the politisation of the streets when the public domain becomes increasingly controlled and regulated through the privatisation of spaces in the modern/global city, especially shopping malls (Jackson 1998). This means that boundaries between the private/public divide become matters of class, cultural, political and social settings and actually express different 'spaces of citizenship' that may change among different classes, races, geographical regions within a country, between urban and rural environments (Sullivan 1995) and between densely populated or dispersed areas.

Summary and conclusions

This chapter made the connection between three basic main issues: first, how different identities create different personal and social experiences resulting in different embodied knowledge; second, how these personal experiences construct the relative, subjective and symbolic meanings of spaces; and third, how these personal and social identities are perceived as excluded or included by different approaches of citizenship in the global city. The following discussions were highlighted:

- *The connection between identities and knowledge* was introduced, pointing out how embedded knowledge is produced, comprehended and internalised differently for people of different identities.

- *Social identities construct spatial knowledge as local* as social identities are constructed out of community's cultural and social norms and values, some of which construct meanings to space, which become the community's local knowledge.

- *Gender identity associates the distinctions between private and public spaces*, which in certain cultures become *forbidden and permitted* spaces.

- *Clothing* is another *cultural construct*, which becomes spatial element and as such affects women's movement in space. This is especially relevant for Muslim women whose culture perceives their bodies as contested sites of cultural representation and therefore the role of clothing becomes crucial in their daily practices.

- *Identities such as age, disability, and sexual preferences* each have their cultural constructions and perceptions of space, which are formulated out of their specific identities.

- *Absolute and relative constructions* of space are another topic discussed in the chapter. Relative perception of space comes out of people's identities and experiences side by side with approaches that perceive space as absolute, as merely physical, containers of both nature and society.

- *Lefebvre's triad of the perceived, the conceived and the lived* can be linked to notions of local and professional knowledge. The perceived means our daily spatial experiences in already planned spaces. The conceived is the space of scientists that is, the planning procedures reflecting professional planning knowledge. The lived relates to the associations of space, the images and symbols. The lived is what formulates our local knowledge based on our daily practices in the perceived.

- Looking at the discourse between *equality and difference* and suggesting new directions of thinking regarding '*citizenship in the global city*' sheds a new light on the debate on citizenship. 'Citizenship in the global city' suggests incorporating notions of comfort, belonging and commitment as part of the citizen rights discourse as will be developed later in the book.

- The connections between *identity issues and constructions of space* highlight notions of *exclusion and inclusion* embedded in the citizenship discourse. These terminologies echo in the gender and cultural discourse especially in the articulation of private and public divides.

Notes

1 Yaniv Abramson (1999) 'Space in the eyes of the blind', a seminar dissertation, Department of Geography, Tel Aviv University.

2 Esther Wolf (1999) 'Culture, space and planning: physical disability in urban spaces', a seminar dissertation, Department of Geography, Tel Aviv University.

3 I wish to thank Haim Yacobi for providing a clear and enriching interpretation of the triad of Lefebvre.

4 This is of course a very problematic issue, as what happens if respect for ethnic minorities' cultural norms means also approving women's discrimination and abuse (see Fenster 1999c).

London and Jerusalem:
Whose planning, whose power, whose diversity?

CHAPTER 4

London and Jerusalem – two cities so different, so contrasting, the global and the holy city. What could be the common base for a discussion on these two cities? It is argued here that they share similar aspects of the power of the 'politics of planning and development' that shape them. In London, the 'politics of planning and development' reflect the capitalist power relations between the different actors involved in the city expansion especially into the East End and Spitalfields–Brick Lane area. The 'politics of planning and development' in Jerusalem mirror nationalistic and territorial interests, using professional tools and planning mechanisms to control and oppress the Palestinian population in the city. In both cases the power relation interests between the different actors is the core issue in understanding the 'planning games' that shape and reshape cityscapes.

> 'I love my city very much, I was born here. I feel very comfortable in the city centre and the city. It is a place where I can find the things that meet my professional and personal needs. At the same time I am bothered by the neglect of beautiful buildings. Every building in the city centre has a very interesting history . . . these buildings need to be restored and to become a part of a new structure.' (Sarit, 50s, Israeli – Jewish, Jerusalem, 27 April 2001)

> 'I love London, it is part of me . . . It is exciting and challenging and I like exciting and challenging things, I like the anonymity, the changing quality, the complexity, it is like a big village for me, I am like a taxi driver; I know my way round physically and culturally. I feel secure in London . . . it's a world city, I came from a small village so I know how to appreciate it.' (Donald, 50s, European – English–middle class, London, 5 August 2001)

Two reflections representing the 'majority-hegemonic' group in each city, the Jewish people in Jerusalem and the white English middle class people in London. What do people of different identity feel towards the two cities?

'There is a problem of transportation which negatively affects women more then men. I am a Palestinian woman and if I want to get to East Jerusalem I can't! There is no direct bus from West to East Jerusalem – lack of transportation is an indication of the policy of apartheid. If I want to get to the Damascus Gate in the Old City which is two minutes from where I live I have to take two buses . . . walking is a problem . . . a city that undergoes a process of deurbanisation marks women as different and excludes them . . . my body becomes a scale, a measurement.' (Aziza, 30s, Palestinian – Palestinian (citizen of Israel), Jerusalem, 7 August 2000)

'London is the heart of capitalism but also the back court of Bangladesh' (Ali, 50s, Bangladeshi–British, London, 27 August 2001)

The experiences of the 'other' – the Palestinian indigenous people in Jerusalem and the Bangladeshi former immigrants in London – are somewhat different to those of the hegemonic dominant group. Aziza's daily efforts to wander in the city become a gendered and national embodied experience of discrimination (see Chapter 8). Ali, a Bangladeshi living in London for the last 30 years, sees the contrasting sides of the city, its capitalist power as a world city but also its localised nature, being the 'back court' of Bangladesh. These are the voices of two marginalised communities that live in the cities, those who struggle for their rights, which are abused by the 'politics in planning and development' in each of the cities.

Like many other cities in the world London and Jerusalem are economically and nationally divided – a very basic spatial expression of economic and social division and power. In London it is the East–West economic/class division; the former known as the poor, immigrant-base area, the latter is perceived as the affluent globalised area. London also has its North–South division: the North indicates the upper–middle-class, white English residential areas, and the South indicates the lower–middle-class, mostly immigrant residential areas. In Jerusalem, the most distinctive, physical and clear-cut division is the East Palestinian–West Jewish national division, but the North–South division also has economic and ethnic meaning: the North is the residential area of the ultra-orthodox Jews and the South is populated by lower middle-class Jewish residents.

In such spatial divisions the nature and functioning of the 'politics of planning and development' is analysed with specific reference to the political, organisational and planning context, highlighting aspects of power relations in the management of the two cities. Finally an in-depth analysis of two case studies is presented. In London, the development proposals for the Brick Lane area in the East End serve as an example of the economic–capitalist power becoming the driving force of the city's expansion towards the East End. In Jerusalem, we look at Isawiye, a Palestinian village located in the outskirts of the city, and examine the story of the power relations involved between the community and the Jerusalem municipality around planning issues.

London: urban planning and global capitalism – whose power, whose knowledge?

The 'politics of planning and development' in London reflect in many ways the capitalist power relations that dictate and control London's cityscapes. The different 'actors' who play major roles in shaping and reshaping urban life in London are presented in this section as related to the analysis of London's planning system and the development projects in the Spitalfields–Brick Lane area.

London – the global city: prospects and challenges[1]

London, the capital of the United Kingdom with about 7.5 million people (12 per cent of the UK population), is considered one of the three leading cities in the world (with New York and Tokyo). Its foreign exchange market is the largest in the world, with daily turnover of around $637 billion, more than New York and Tokyo combined. It is the world's largest fund management centre, with nearly $2.5 trillion worth of institutional holdings in 1999.

It is also the world's largest international insurance market, with gross premium income of £14 billion in 1999. London's GDP is £180 billion a year, larger than many of the EU countries and with GDP per capita exceeded only by that of the United States and Luxembourg. Finance and business services account for around 40 per cent of its GDP with 300,000 people employed. There are 481 banks in London, more than any other centre worldwide. Its economy has grown faster than the rest of the United Kingdom for most of the last 15 years and its contribution to the national economy has grown equally. London makes a major contribution to the economic prosperity of the United Kingdom by providing over 4 million jobs in the rest of the country, which depend on London's demands for goods and services. Of the top 500 global companies 375 have London offices. The city is considered a world cultural production centre. The cultural and creative industries have an annual turnover of £25–29 million with 700,000 people employed, just fewer than 20 per cent of the city's workforce; 800,000 people commute into London to work. It is a city of cultural diversity. The English language, the closest to an international medium of communication, is spoken along with some 300 other languages. It has resident communities of more than 10,000 people from each of 34 different countries, with one-third of its population coming from Black, Asian and other ethnic minority communities. It is a city of tourist attraction. It receives around 30 million visitors a year, who spend £9 billion annually.

However, side by side with its impressive growth the city's economic, social and infrastructure systems are approaching crisis, which in many ways is characteristic of global cities. London has the highest percentage of high disposable income households in the country but also the highest percentage of low disposable income households. In November 2000 over 1 million people in London (13 per cent of its population) were living in households dependent on income support. London has 20 of the 88 most deprived local authority districts in England and 16 of them rank in the top 50 on any of the six indices of deprivation measures. Of the children in London 43 per cent live in households below the poverty line once housing costs are taken into account; 29 per cent of pensioners in London live in households below the poverty line once housing cost are taken into account and 46 per cent of the children in inner London secondary schools are entitled to free school meals.

London has more unemployment than Scotland and Northern Ireland together and its level in the inner city is twice the national average. Inner London East has one of the highest areas of unemployment in the United Kingdom: 13.8 per cent. Its average unemployment is 6.9 per cent as compared to a national rate of 5.2 per cent. These unemployment rates coexist alongside a dire shortage of skilled workers. Nor is it the case that the prime shortages are of senior financial analysts, surgeons or lawyers. By far the most significant shortages are of skilled manual and office workers. In addition there are acute shortages of nurses, teachers, police and other middle-ranking public sector workers, mostly because of the high cost of living, of housing and transportation in London.

The housing situation in London is another potential crisis. The city has an acute shortage of housing, both for rent and for purchase. There were over 50,000 homeless households in London in temporary accommodation at the end of March 2001 and many needing subsidised homes to rent face years on waiting lists. The price of housing has climbed as high as 81 per cent more than the national average. As a result 85,000 London households cannot afford their mortgage payments and 178,000 households in the private rented sector cannot afford their rent unassisted. Consequently employers in the public and private sectors are finding it increasingly hard to recruit and retain the skilled staff they need. This is why teacher vacancy rates are 3.5 times higher than in other English regions.

Because of its growth and the growing number of people coming to the city – 1.1 million people travel into central London during the peak period with over 3 million tube journeys a day (for working purposes or tourism) – its transportation system is about to collapse, failing to meet the needs of its customers. Its disastrous situation is such that (before the introduction of congestion charging) average daytime traffic speeds in central London are 10 mph, which is no faster than in Victorian times, mostly because some 700,000 vehicles drive into the centre of London every day.

In many parts of the city the education system fails to meet the necessary standards. Because of its cultural diversity, for more than half of the pupils in inner London English is a second language and is not the main language spoken at home. Of secondary school pupils 21 per cent are unable to speak fluent English. The shortage in teachers that has already been mentioned is one of the causes of the low level of success of pupils in London. Inner London students leave school with significantly lower qualifications than outer London and the United Kingdom as a whole.

The health situation in London is also deteriorating. Londoners suffer disproportionately from respiratory illness, and independent estimates suggest that 350–400 premature deaths a year are caused by particulate and sulphur dioxide emissions from vehicles. Traffic is responsible for 90 per cent of air pollution in London.

London is a city of crime. More than 10,000 incidents of racial violence and abuse are reported to the Metropolitan Police every year. In 2000 there were 155,000 offences involving violence. The proportion of total reported crimes in the United Kingdom has increased from 16.8 per cent in 1995/6 to 20.7 per cent in 2000. At the same time London has far fewer police per capita than New York. There is one police officer for every 161 citizens in New York whereas in London the figure exceeds 286. For every police officer in New York there are 7 crimes, in London there are 28.

What makes London a mixture of insightful opportunities and harsh daily life? How can the city be a place of so many possibilities and chances but be so harsh in its basic services and infrastructure? Let us look at one major aspect of London's city life – its planning system – in order to find out more about these contradictions.

The planning system in the United Kingdom and in London[2]

Central government – Local Governance in London

London, the national capital, is directly subordinated to central government – a programmatic, nationally controlled, multiparty system structures its mode of political representation, and its municipal government is decentralised to 33 local governments:

> Organised like the national Parliament, local councils are run by the majority party with an executive group consisting of a leader and chairs of the various functional areas for which the council is responsible (for example education, housing, social services, finance, planning and transport). Councillors are elected by ward; the number of councillors for each borough varies. Under London's governmental system not only does a representative body operate at a level far closer to the neighbourhood, but the location of administrative offices in every borough allows its citizens to carry out their transactions with government much closer to home than for example in Jerusalem. (Fainstein 1994: 82)

In the United Kingdom there is a direct link between the government and the local councils' policies. This is expressed in the fact that centralised parliamentary government has consistently limited the autonomy of local government. The principles of cabinet rule and party control make national party programmes dominant both through the mechanism of parliamentary supremacy and, for the national opposition parties, through the party machinery. Similarly to Israel, the final responsibility for local planning rests with the cabinet minister, in the United Kingdom it is the head of the Department for Environment, Food and Rural Affairs and in Israel it is the Ministry of Interior. In the United Kingdom as much as in Israel the power of central government to curtail the level of local taxation and expenditure further constrains local authorities.

Financially, however, local authorities seem to act independently. They raise most of their funds themselves – until recently rates (that is, property taxes on households and businesses) along with fees, rents and property sales were their principal sources of revenue. In 1990 rates were replaced by 'poll taxes', which obliged each adult householder to pay the same amount, not considering income or property holdings. In 1993 a new form of local taxation called the 'council tax' reintroduced a tax based on property values.

The power relations between central government and the London government have shifted according to the political ideology dominant in central government, with Labour advocating decentralisation and the Conservatives working with greater central control. During the 1970s the British Labour Government formulated an inner-city strategy modelled on the War on Poverty and Model Cities programmes. Under the leadership of Ken Livingstone, the Greater London Council (GLC) of the 1980s sought to combine an industry-based economic development strategy with political radicalism (Fainstein 1994). The GLC tried to combine elaborate social service programmes with community participation and economic revitalisation strategies. However the GLC was abolished under the Thatcher government in 1986 because it was considered too Labour oriented in its policies and development orientations, despite the referendum showing support for its continuation by an overwhelming majority of Londoners. The conflict over the abolition of the GLC had importance beyond its immediate implications for planning and administration. Under Ken Livingstone the GLC had become a significant political force in opposition to Thatcherism. Its commitment to low transport fares, manufacturing investment, social housing and 'fringe' cultural institutions brought upon it the wrath of the Conservative central government. During the six-year period from 1978 to 1984 its expenditure increased in real terms by 65 per cent. Its abolition was widely and perhaps correctly interpreted as an attack on municipal radicalism and local autonomy. However, even when the GLC functioned at its highest, it did not possess as much authority as local government in, say, New York City (Fainstein 1994). The central government in the United Kingdom could always exercise authority over the GLC's decisions, and the borough authorities kept control of most housing provision, land use planning and housekeeping functions such as street cleaning and fire fighting. Furthermore, the leader of the GLC did not have the dominant position in framing the city's agenda.

Thus, instead of social-oriented policies, which the GLC aimed to promote, the Conservatives who came into power used real estate development (in contrast to job training or infrastructure investment) as their primary strategy for stimulating expansion. They identified global city status as the hallmark of their economic advantage and fostered those forms of development – especially first-class office space and luxury housing – that responded to the needs of the upper-class financial and advanced service industries participating in world economic coordination. Consequently, in the London of the 1980s, expropriation became a real threat to all working-class communities. (Forman 1989). Forman uses the terminology 'expropriation' to illustrate the negative effects of such policies on the public, especially working-class people. The London version of expropriation is about forcing up rents, taking away security of tenure and breaking up council estates. This has happened in many areas becoming regenerated or gentrified within the boundaries of central London (Forman, 1989). This perhaps explains the closely related connection between the city's capitalist development and expansion and its effects on the deterioration of quality of life for middle and lower-income people.

Once London lost its metropolitan government in 1986 the power of central government increased. Overall executive authority shifted to the Department of the Environment (DoE) and ultimately to the Prime Minister, although the DoE's London office has never devised a policy for the city.

With GLC support gone, local authorities which objected to central government's policies lost access to resources that had helped them in finding a more independent path. Also a local political paralysis was meant to be produced: the Conservatives simply wiped out a whole level of government inhibited local government activism by simply showing that any local deviation from central government policy would have little chance of success. By the end of the 1980s local authority politicians who opposed central government policies had given up and went for market-led development. Thornley describes the city thus (1992: 2):

> London ever since the demise of the GLC has had no single authority. There has been no voice expressing the concerns for London as a whole, extolling its image or dealing with its strategic problems. Although no one is advocating a return to the large bureaucracy of the GLC there is widespread feeling, from developers to local communities, that there is a need for some kind of body to play this role.

Labour came back into power in 1997 after 18 years of Conservative rule. The effects on the planning system and local authorities–government relationships took a few years to realise. In 2000 the Greater London Authority (GLA) was established following the Greater London Authority Act of 1999. The GLA is one of three assemblies to be established in the United Kingdom, the others are in Wales and Scotland. The GLA is a new strategic city-wide government for London. It covers the 32 London boroughs and the Corporation of London. It is made up of a directly elected Mayor – the Mayor of London – and a separately elected Assembly – the London Assembly. The Mayor is responsible for strategic planning in London. He is in charge of producing a Spatial Development Strategy for London, which has been named the London Plan. This plan sets out the challenges, visions and principles of London's development up to the year 2016 with the main aim of retaining its position as 'the leading city in Europe and one of the three leading cities in the world' (London Plan, May 2001). The draft of this plan has been publicised for consultation on a special website and is still on the site as initial proposals in 2003. The London Plan will replace the existing strategic planning guidance. Once published, it becomes the plan that the London boroughs' unitary development plans should confirm to.

In addition, the London Development Agency (LDA) has been established in order to formulate, promote and deliver the Mayor's economic development and regeneration strategy for London. The LDA was established in July 2000 and joins the eight Regional Development Agencies already set up in the United Kingdom. In fact this new structure adds another hierarchical level between the government and local authorities. The LDA power derives from the Regional Development Agencies Act of 1998 as amended by the Greater London Authority Act 1999. One of the major expressions of the decentralisation process happening in London is that the Mayor of London and not the Secretary of State for the Environment is answerable to the LDA. The LDA may be viewed as a replacement of the GLC but with much wider authority regarding planning and development in the city.

These current developments in local and central government provide grounds for optimism as to the chances of changing London's current difficulties. However, it is too early yet to evaluate these prospects.

The recent history of planning in London

Patrick Abercrombie's plan for London after World War II, updated by the Greater London Development Plan of 1968, set out the criteria for green-belt preservation, peripheral manufacturing and population dispersal, thereby providing a framework for development. These principles have been used also in British colonies and protectorates such as Jerusalem.

> Planning officials emphasised clustered development around town centres and a sharp demarcation between city and country; regulation was much stricter than for example in New York and the preservation of the green belt around Greater London constituted a cardinal principle. Preservationist concerns played a major role in limiting and directing development. London still has more restrictions on the right to build than New York. (Fainstein 1994: 83)

Each local authority within the region produces development plans to guide growth within its jurisdiction. Until the 1970s central government policy was to decentralise population and economic activity out of London but this was eventually reversed as London began to lose industry along with population at an unanticipated rate. In the mid 1970s the Labour government encouraged central London's planners to seek to retain population and industry in the city. Thatcher's emphasis on using the market to allocate investment meant a notable relaxation of planning controls during her time in office.

The end of the GLC in the mid 1980s meant the elimination of any authoritative planning body for the London region as a whole and the fulfilment of the Thatcher government planning philosophy, which was to depend and rely on 'the initiative of the private sector and individual citizens and effective co-operation between the public and private sectors, not an imposition of a master plan' (www.london.gov.uk/mayor/case_for_london). In the absence of a planning authority for the metropolis the London Planning Advisory Committee was set up to give planning advice concerning London to the government. It consisted of a small staff and a joint committee of 33 members drawn from London's boroughs. But it was considered toothless. The DoE was responsible for planning the metropolitan region but it did not see its mission as the forceful coordination of local initiatives or the preparation of a detailed comprehensive plan. It acted primarily as an appeals body, frequently overturning lower-level decisions that had rejected planning permission to developers. Borough councils became more and more inhibited from rejecting development proposals for fear of having to defend their decisions on appeal to the Secretary of State.

Fainstein (1994) writes that London in the 1970s presented a model of greater state intervention mainly because of the political capacity of the government to establish a sphere of autonomy for itself: the existence of the dominant class of capitalists that interpreted its interests collectively and held a paternalistic vision of its role in society and the political capacity of subordinate classes to influence state policy in their own interests.

But in spite of this government dominance in the 1970s local interests continued to affect London's fate during the Thatcher–Major era. Labour control over half of the boroughs meant that opponents of Conservative policies could shape local authority agendas and use them as bases of resistance to the initiatives of developers. Labour councils used to be dominated by white, male, trade union leaders but this composition changed during the Thatcher period with new kinds of activists, who were even less favourable to development proposals. Blacks and feminists successfully challenged the white male dominance of Labour strongholds in East and South London. Also Asian and Afro-Caribbeans achieved representation on councils in boroughs with large minority populations, and women also gained additional council seats, but nevertheless class still defined the major fissures in London society and politics.

To conclude, for much of this century urban planning in London concentrated on controlling and improving the physical environment. Planners engaged primarily in determining land uses and proposing public investments in infrastructure, amenities and housing. The premise of planning was essentially a determinist–economic response. Holding on to a supply-oriented view of urban space, planners expected that private investors would avail themselves of adequately serviced, centrally located land without further incentives. As analysed in the next section, this situation changed during the 1970s and 1980s when private developers became dominant actors in the planning arena.

Actors involved in the planning apparatus in London

As mentioned above, until the 1970s urban planning's justification in London lay in its comprehensiveness, an orientation to the long term, protection of the environment, and preservation of the public interest through orderly development and attention to the interests of all social groups. This approach was largely criticised, as shown in Chapter 2. Within this background the discourse in which planners in London interpret the world and communicate their intentions has shifted from long-term concerns with environmental quality to an emphasis on short-term accomplishments.

After the dominance of the rational comprehensive approach during the 1960s and 1970s a different but broadly generalised planning mode was emerging. Rather than framing and testing an exhaustive and abstractly constructed set of alternatives, planners concerned with maintaining neighbourhoods or bringing new investment into the central business district aimed at discerning targets of opportunity. Instead of picturing an end state and elaborating the means to arrive at it, they established narrow goals and went for a mixture of devices, which will permit at least some forward progress.

But even planners working for government or non-profit groups who do focus on distributional effects are absorbed into the discourse of business negotiations and a single-minded concentration on the deal at hand. Planners began to speak in the same language as investment bankers, property brokers and budget analysts. The old argument for planning comprehensiveness and reducing negative effects on the environment has changed to one of competitiveness and efficiency (Fainstein 1994). Planning has become directed at achieving marketability of its product – urban space. It must be emphasised that many planners still work at traditional planning functions such as establishing zoning regulations or transport programmes, and this new discourse in planning affects them little. But for planners concerned with economic development the focus has changed radically: they perform a mediating function, bringing together private investors and public sponsors. Negotiation rather than plan making has become the planner's most important activity.

The resulting landscape of these processes reflects the piecemeal, accommodationist mode in which planning is carried out today. Most office projects and upper-income condominiums, which once would have formed part of a larger development programme, are now single-site efforts uncoordinated with their surroundings, contributing to the postmodern vista of chequer-board development, un-matching architecture, uncontrolled congestion, and sharp juxtaposition of rich and poor. (Fainstein 1994: 101)

In London, planners seek to persuade private developers to build mixed-income projects in which market-rate units subsidise the much smaller per-centage of low-income ones.

In the remainder of this section two of the significant mechanisms of London's planning system – public–private partnerships and the planning gains and exactions mechanism – will be discussed, highlighting its capitalist-oriented nature and perhaps providing another explanation of the socio-economic contrasts embedded in London's city life.

Public–private partnership

The common interest of both government and developers in maximising the potential return on a site allows public–private partnerships (PPP) to flourish.

> Most of these arrangements involve joint public–private participation in major commercial projects, but some agencies assisting low-income households and neighbourhood businesses, as well as some Labour-dominated borough authorities, have also resorted to this method of resource mobilisation. (Fainstein 1994: 108)

These relationships characterised planning in the mid 1970s and have their roots in changes in global relations that changed urban economies and increased inter-urban conflicts. The pressure for reconfiguration of the existing built environment grew and became the basis on which the governing regimes of London built their growth strategies. Land development provided private entrepreneurs with opportunities for massive profit; facilitating such development gave planners power over the private sector. Governmental agencies had largely stepped down from their earlier role in urban regeneration, whereby they gained control and serviced land and built public facilities. There was a mutual need of public and private enterprises for facilitating lower development costs and thereby making projects more attractive to investors as most of those projects could not be developed without the provision of infrastructure. This need for governmental intervention created the opening for bargaining, in which public officials traded governmental approval or capital spending for private contributions to housing or public facilities. These tendencies called new attention to the idea of the 'public interest' that official planners are supposed to promote as part of their duties. Planners are especially challenged by the 'public interest' when trading planning gains and exactions.

Planning gains and exactions

Trading public and private benefits is implemented through the planning gain, which has constituted the principal vehicle by which the urban redevelopment process proceeded in London during the 1980s. These arrangements took a variety of forms including, among others, the trading of planning permission in return for the construction of affordable housing, the collaboration of public agencies with community organisations in low-income neighbourhoods to form local development corporations, and the granting of tax holidays to private investors within enterprise zones in the hope of employment expansion.

Public officials in London have expanded the concept of exactions beyond its original meaning. More recently planners have made developers contribute to off-site improvements, including affordable housing and day care centres, as a condition for approval of their proposals. This is possible under Section 52 of the 1971 Planning Act, which gives councils the power to oblige developers to pay some 'community benefit' along with their commercial scheme. Councils and planners have been justifying offices by bringing up this 'planning gain' (Forman 1989). Developers anxious to get their projects moving will frequently offer a range of concessions in order to elicit cooperation from planners who can no longer look to higher levels of government for substantial capital assistance and thus see negotiations with developers as their principal opportunity for obtaining community benefits.

The package of public permissions and developer obligations is incorporated into a development agreement. Because British local authorities have great discretion in the granting of planning permission, a London local authority planner occupies a potentially dominating position in the quest for planning gain. Planning gains are 'ad hoc' arrangements, but the formal vehicle through which government entices private developers to participate in fulfilling their economic development objectives is the Urban Development Corporation (UDC). It operates much like a private firm, employing the entrepreneurial styles and professional image-building techniques more customary in the corporate than in the governmental world. After 1979, when Thatcher became prime minister, the UDC became much more central to urban policy making than previously.

Reading through the analysis of London's politics of planning and development we can see the shift in the last decades from public-oriented policies to more capitalist-oriented policies. This shift is perhaps one of the explanations of London's mixed character of being both a city of insightful opportunities and a place of harsh life. The following case study emphasises these contradictory forces in shaping London's cityscapes at a much more localised level of planning.

Capitalist power and development in the East End – Tower Hamlets, Brick Lane area[3]

The East End has been known for centuries as a first stop for waves of immigrants entering England, from the Huguenot silk weavers at the end of the nineteenth century through the Irish and Jewish immigrants of the beginning of the twentieth century to the recent Bangladeshi immigrants moving into the area in the mid-1960s.

In the 1970s global economic activities' demands for office space grew around the City of London. The City, the heart of the capitalist action and activities, had perhaps other directions to expand than the East End, such as towards Fleet Street or to the West and South (see Figures 4.1 and 4.2), but as Forman mentions (1989: 142): 'East is best, its both the nearest and the cheapest'. A very negative move, in his eyes, for the residents of the area for: 'there is little the City has to offer the people in Spitalfields – the more money there is invested in the area, the poorer its residents become' (p. 141). Forman sees the recent developments in the area as a battle between the city's expansion, the developer's motivations and the local voices of the poor. Most of Spitalfields' residents are poor as this ward is ranked the most deprived ward in Tower Hamlets, itself the most deprived borough in London.

Spitalfields is the economic and social 'Other' of the City of London' (Jacobs 1996: 70). The 'Other' in this case consists mostly (80 per cent of population) of Bengalis living and working in the Brick Lane area. The battles over development in the area serve as an example of how the planning and development machinery becomes a showground of power relations in which hegemony and dominance shape the future of city spaces. In the Brick Lane area the echoes of these battles resonate in two major development projects. The one famous for its big opposition and objections is the Spitalfields Market project, the other is the Bishopsgate Goodsyard project. But before going into the details on these projects some background information about the local politics and planning is presented.

Tower Hamlets and central government sponsorship – the City Challenge and the Single Regeneration Budget schemes[4]

The Liberal Democrats ruled Tower Hamlets' council from 1986 until 1994, a period considered as dominated by racial violations. Labour came into power in 1994 and stayed in the council until 2002 when new elections for the local authority's council took place. The local leadership in the council is still dominated by white people; only 22 of the 50 elected councillors are Bangladeshi (44 per cent). But in many ways this is an achievement, since only in the 1970s when Tower Hamlets borough council passed from Labour to Liberal Democrat control the area had experienced waves of racism and violence against Asians. These actually pushed the Bengalis to become more politicised and politically involved. With the growing polarisation of the East Asian community, its representation in Tower Hamlets borough council grew from 9 out of 50 representatives in 1992 to the 22 councillors in 2000 with good prospects of increasing their numbers in future elections.

Since Tower Hamlets is one of the poorest boroughs in the United Kingdom – 17 wards (out of 19) have been included in the 88 most deprived wards in

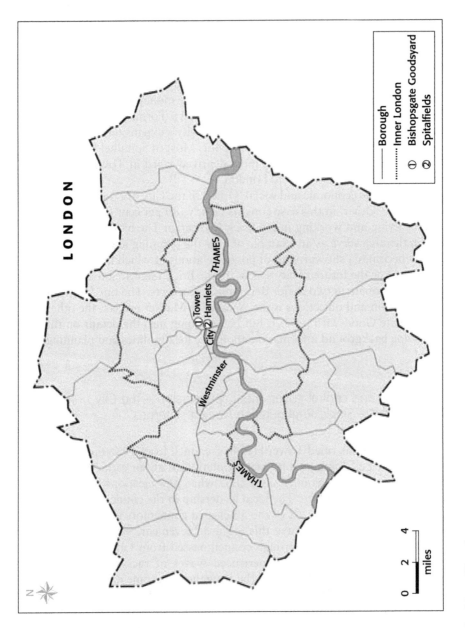

Figure 4.1 Map of London

Figure 4.2   The City of London from the Spitalfields Market site

the country – it receives a government budget to deal with issues of equality and social exclusion. The council worked on two main themes: regeneration and equality. The regeneration budget is allocated to three different areas within the authority's boundaries interchangeably so that each year another area receives a portion of the budget. The last three years were dedicated to working on economic initiations and recently the council worked on community empowerment, emphasising 'self-empowerment' namely building on local people's organisational skills.

Another major regeneration effort in Tower Hamlets sponsored by central government is the City Challenge. This has been established as a new mode of planning whereby decision making on budget allocation and projects takes place in an agency outside the regular governmental structure. The City Challenge grant programme was intended to stimulate activity by the kinds of elite corporate entities and community development groups long active in US inner cities, but not yet visible within the United Kingdom. In response to a request for proposals, Tower Hamlets council applied for a City Challenge grant for Bethnal Green, one of its neighbourhoods. It received the grant in 1992 for community-oriented projects. The government contribution was intended to leverage private sector commitments, and community organisations were supposed to assist small-scale entrepreneurship.

The City Challenge framework functioned until 1995 as a replacement of the Task Force system, which operated until 1992. From 1995 the budget for regeneration projects was transferred to a different structure, more regional than sector oriented – the Single Regeneration Budget (SRB) for Environment, Transport and the Regions. Recommendations for the use of this budget are within the responsibility of the Regional Development Agencies or the London Development Agency. These procedures are intended to ensure greater coordination between the different sectors involved in regional regeneration projects. Local authorities bid for the money in two phases; the first is an official one in which they have to complete forms. In the second phase they have to submit a regeneration plan for their area for the coming one to seven years. So far six SRB schemes (from 1995–2000) have been provided for local authorities in London.

These schemes represent government efforts to diminish the negative effects of global expansion. The government has been working on two main regeneration paths; the first, on social exclusion, established a special unit, which published some 18 reports on social exclusion in different parts of the United Kingdom, and a national report. For the 88 neighbourhoods or wards that were ranked as the most deprived in the country the government set up a scheme called 'neighbourhood renewal strategies' to focus on specific local problems connected to social exclusion. The second path to deal with regeneration is the modernisation process, which is connected to issues of equity. A special body has been established, the Local Strategy Partnership (LSP), with the intention of setting up such bodies in every local authority so that modernisation is implemented with equality. LSP has worked out the internal reorganisation of modernisation, which functions in two central themes; the first is to establish a community plan that will set up the goals and the strategies of modernisation and the second is to work on the regeneration aspects of the neighbourhood renewal strategies. These official activities clearly demonstrate central government efforts in the last decade not only to decentralise and equalise development but also to support local communities in their struggle to maintain a good quality of life in their neighbourhoods. These schemes could also be seen as another indirect mechanism to ensure the balance between global capital interests and local communities' interests. Central government and the GLA are interested in both global capital expansion and local communities' interests as part of their political obligations to their citizens. It seems that the regeneration schemes reflect an effort to bridge what is sometimes seen as a gap between these two sets of interests.

Only time will tell whether these organisational and bureaucratic changes will allow more equality and a less capitalist-oriented planning system to ease the everyday life of the city's citizens.

The Bangladeshi community

The Bangladeshi people are the 'Other' within the city of London. They arrived in the Brick Lane area mostly in the 1970s from the Sylhet district because of their historical connection, their ancestors having worked on British imperial ships in the nineteenth and twentieth century (Forman 1989). They come from rural areas, living in villages based on an extended family system that could include 100–200 people living in spatial proximity that reflects their social family ties. This pattern of extended family structure would be adopted later in their spatial housing arrangements in the Brick Lane area. Village loyalties ensured that once someone had got a foothold in London they would do what they could to help others from the same village or area in Bangladesh.

It should be emphasised that most people who arrived saw their stay as temporary. They were here for their families, were sending money back to them regularly, and were still taking direction from the head of the family back in Bangladesh (Forman 1989). In 1981 there were some 9,800 Bangladeshi-born people living in Tower Hamlets, in 1984 there were 15,000 and in 2000 there were 60,000, which is some 20 per cent of the total Bengali population living in the United Kingdom and about 40 per cent of the total population of the borough (Kamaluddin 2000). The first waves of Bengali were men. Only after a few years when their plans to return were constantly delayed did they bring their families. Those who started to arrive in the 1970s faced enormous problems in adjusting to the new way of life in London and faced hardship and terrible housing conditions. Women faced even greater difficulties in adjusting as they had lost the family support system that was the basis of their life in their villages. In London in spite of the housing proximity this was not the case. With no knowledge of the English language, and responsibility for establishing and maintaining a household, they had to adjust quickly to housing conditions and spatial restrictions strange to them. Today the Bangladeshis are the poorest community in the country with regard to health and education. About 70 per cent of Bangladeshis cannot speak or read English, and 36 per cent are jobless (Kamaluddin 2000).

The Bengali population works mostly in manufacturing and the garment clothing industry but also in restaurants and groceries, sari shops and Bengali supermarkets (see Figures 4.3 and 4.4). Spitalfields ward has been renamed 'Banglatown', which is how it appeared in the elections for the borough council in 2002. Banglatown in the last decade or two has become a place of commercial and cultural life full of restaurants and bazaar areas in order to attract tourists but also as the reinvention of another place (Jacobs 1996). The promotion of Banglatown by the Bengali population is analysed by Jacobs as: 'an activation of an essentialised identity category by one sector of the Bengali community within the terms of the enterprise-linked development opportunities available' (p. 100), that is, the adaptation of some of the Bengali community members to the economic prospects of using their ethnic identity to promote their economic prospects. Jacobs also notes the ideological differences between Bengali businessmen, who are interested in promoting the economic idea of Banglatown and perhaps gaining from the development of the area, and some of the local left, who see the battles against the developers' intentions in the area as uncompromisable.

Three generations of Bengali people live today in London, each with different attitudes and political involvement typical to immigrant society (Nisar

Figure 4.3   Bengali supermarket, Brick Lane, London

Figure 4.4   Sale in a sari shop in Brick Lane, London

Ahmend, 28 August 2001). The first generation of immigrants arrived in the early to mid-1970s when they were in their mid-30s and early 40s. They are considered the old timers, the first generation to come that was never involved in political activism. They are now mostly retired and in many cases are dependent on their children.

The second generation consists of those who arrived in the 1970s as children, are now in their mid 40s, and are the middle generation in the Bengali community. They became politically active in response to waves of racism arising in the area in the 1970s and the 1980s. They became involved in anti-racist movements, in trade union activities and in Labour Party political activism. They are people with split identities. They have lived here most of their adult life but were not born in the United Kingdom. They are what one of their leading activists calls 'the schizophrenic generation'.

The third generation consists of Bengali who were born in London. They represent the modernised and educated generation. Some are very politicised and some others choose to practise the Islamic religion as a way of life. This generation had a much clearer 'Bengali' identity then merely 'Asian'. For most of them 'home' is Bangladesh, although they do not wish to nor can they go to live there. They posses split identities too. As one of them said: '*We'll never be English men as much as they are but we are British in terms of our education, mannerisms and thoughts . . . we can't live anywhere else*' (Gaziul, London, 10 August 2001).

As a result of their political activism and with the help of government regeneration schemes and some left radical British whites a large number of community groups were established in the area, gaining growing influence in local politics. These groups fought both against racial attacks and for obtaining decent living accommodation for the community's members. When developers began to show an interest in the area during the 1980s these groups started to be involved in active opposition to the developers' intentions. While in the Spitalfields Market project they were involved only to a minor extent, in the Bishopsgate Goodsyard project they were more involved in the planning gains and exactions. However, it should be noted that their reactions and attitudes to the development projects in their area were mostly ones of acceptance. As one of the leading Bengali persons in the area told me, they had three options: either to say 'no' to the developers as the left radical whites had done in their battle for Spitalfields, or they could have said 'yes' to the developers like the people of the Spitalfields Historic Building Trust (who said 'yes' because they were concerned about the 230 Georgian houses in the area, Forman 1989).[5] But the Bengalis knew they couldn't get much because they lacked the money and the power to negotiate with the developers, so they considered the third option, which has been to say 'yes – but' that is, to negotiate for community benefits from the development project. As seen later, in the Bishopsgate Goodsyard project, they chose this third option to negotiate over their planning gains.

The plot – capitalist power in the Spitalfields project

Spitalfields Wholesale Fruit and Vegetable Market met the demand in the area for 300 years until it was removed to another site in Hackney. Its 11-acre site is now partly home to an indoor market and waiting to be replaced (at least part of it) by a multi-storey office building providing space and services for the expanding City. Its geographical proximity to the Broadgate complex in Liverpool Street makes this site appealing to developers (see Fig. 4.1).

In 1987 the City of London Corporation, which owned the Spitalfields Market site, issued a tender document for the site. The Spitalfields Development Group successfully proposed a multi-use, predominantly office scheme for the market buildings and their surroundings. As the first step it expended some £50 million to move the market to Hackney, which needed a special Act of Parliament to revoke the royal charter that had established the market in 1600 (Fainstein 1994). In the two years it took to get it growing objections to the project arose especially from the local activists, 'Save the Spitalfields Campaign'. Ben Thompson, an American architect with a British firm as a local agent, then modified the plan. His modifications overcame the conservationists' objections, but Labour politicians continued to fault the scheme for its alleged insensitivity to local residents. Planning permission was granted in 1992 by Tower Hamlets local authority but in 1993 the DoE was still considering the proposal. The plan as approved by Tower Hamlets called for 1.1 million square feet of offices and a 16-storey modern glass tower, designed by Sir Norman Foster (Jacobs 1996).

Within the Spitalfields Market development area private–public Partnership and planning gains and exactions find their clear expressions in the politics of planning and development, which shape and reshape the area. The development consortium interested in the area donated 12 acres to a community land trust so as to elicit cooperation from neighbourhood groups; the community is further bargaining for the commitment of £40 million for community development efforts, 127 units of housing, and a job-training scheme. It also agreed to the payment of a £5 million contribution to a charitable trust and to a pledge of £150,000 per year for job training.

Forman (1989) criticises these 'planning gains', arguing that if what the planners gained in these negotiations is needed for the welfare of the people living in the area, the land should be zoned to provide the community with the services it needs and service development not squeezed on to the edge of the office development area. 'The land', he argues, 'won't accommodate both' (p. 158). People should live there instead of offices being built. Moreover, argues Forman, if the local council were interested in using the land to benefit the local working class people would be consulted over its use, which never happened in the past.

In early May 2001 the developers started knocking down almost half of the site. Given the fierce fighting against the demolition that has lasted for the last 15 years the developers are doing their best to sweeten the sour taste of the

Figure 4.5   Spitalfields Regeneration Programme

reality of the market being knocked down. Their PR strategy is to make this project look as if it is almost the best thing that has happened to local communities. To be more precise, they present a shopping list of the 'goodies' that local communities get out of this 'regeneration' programme on big signs located at the entrances of what remains of the market (see Figure 4.5). These signs detail the 'highlights' of the project in terms of the amount of money spent:

- Construction and functioning of the SPITALFIELDS MARKET in Stratford – cost £50,000,000.
- Grant of £5,000,000 to Spitalfields Market Community Trust.
- Grant of £750,000 for Specialist Community Training.
- £13,000,000 towards 118 Social Houses and Leisure Centres in Tower Hamlets.
- Environmental improvements – £1,000,000.
- Two new community centres in Commercial Street.
- Retention of the stall market in the saved Old Spitalfields Market.
- Key private sector partner for community-based regeneration programmes.

On the bottom of this list it is written in capital letters: 'WORKING WITH LOCAL COMMUNITIES'.

Out of this 'shopping list' all that materialised was the grant to the Spitalfields Market Community Trust and the money donated to build the social
houses. The rest of the 'highlights' never materialised or not as they should
have done. For example, the two community centres in Commercial Street are
actually only two small offices, which hardly function (Ahmed, 28 August 2001).

At the time of writing (2002) work is carried out on only half the site but the
developers have planning permission to develop the whole site. With the coming recession it seems that one half of the market is saved for the moment
because capitalist power has slowed down but it is probably only a matter of
time before the whole site is knocked down.

The whole idea of knocking down old buildings is very much against the
idea of 'old and new' that Sir Norman Foster declared at the exhibition dedicated to his work in the British Museum in the summer of 2001. Sir Norman
himself implemented this principle in both the Shackler Galleries at the Royal
Academy of Art and the British Museum Great Court, and it was also implemented in the redesign of the Tate Modern on the site of an old power station.
But the point seems to be lost in the Spitalfields Project where all or part of it
will eventually be knocked down and replaced by an office block. Thus this
project becomes one of the bitter emphases in the dominant role of capitalist
power that, with the aid of planning and development policies, in fact dehu-
manises urban life in the city. ˙

The Bishopsgate Goodsyard project

The project occupies two adjoining sites comprising 27 acres – the Truman
Brewery and the Bishopsgate Goodsyard. The Truman Brewery, owned by
Grand Metropolitan PLC[6] has been shut down since February 1989. The
Bishopsgate Goodsyard was formerly a railway station and goods yard. British
Rail, the owner of the Bishopsgate Goodsyard, and Grand Metropolitan PLC
saw the disused sites as yet another opportunity to transform their property
holdings into negotiable assets. Grand Metropolitan PLC and London and
Edinburgh Trust PLC successfully competed to develop the two sites, mainly
for office use.

Community groups, which opposed these developments, organised themselves in advance to act in a more efficient way than in the case of Spitalfields
or the Docklands. They established an umbrella organisation called Spitalfields
Community Development Group (SCDG) to work together with the developers to produce a Community Plan. The SCDG included a local Bengali think
tank that was established to become more involved in actions and negotiations
regarding this plan.

The developers adopted a 'strategy of dialogue with the local community'
from the outset. They established a Community Development Trust, which
was a partnership between the landowners, the community and the local planning authorities. The developers were aware that this partnership approach

broke new ground. They even labelled it a 'pioneering partnership' between developers and community (Jacobs 1996). In return for the community agreement for office use they focused on two major requests, jobs and training and proper housing. A joint committee of the developers and the community representatives was set up, initiated by Task Force and by Business in the Community, both semi-governmental organisations established to assist local initiatives. After a series of meetings in the mid-1980s between the developers and the community representatives the offer that the developers suggested was for the transference of the 12 acres of land East of Brick Lane to the Community Development Trust and a donation of money to provide a training centre, sports halls, nursery and job agency and the provision of on-site training (Spitalfields Community Development Plan 1989–90). Unlike the Spitalfields case here the community representatives were involved in the negotiation. Nevertheless the outcome raised many objections, mainly from the white left activists who argued that the community plan was similar to the developers' plan so that the community gains were very minor. In spite of this criticism the Bangladeshis chose to carry on with the negotiations. The Bengalis argued that the criticisms from the elite English were neither practical nor ethical. No one, they argued, could stop the development, which would eventually take place, so they thought that they had better benefit from the gains. In the end the project did not materialise because of the collapse of the economic market in the early 1990s. In 2001, the Bishopsgate Goodsyard office development project does not seem likely to be realized. In the meantime another developer bought the Grand Metropolitan Brewery site and it is used as an arts and cultural space. The Goodsyard site is intended to be used as an artists' market.

The two case studies show how the dynamics between the different actors involved in the planning process ultimately serve the interests of the developers and indirectly the interests of the municipality. The Bengali people seem to be the most realistic. They understand the power of capitalism and realise that the best strategy is to make sure that they gain from the development in their neighbourhood. Indeed community involvement is much more impressive in these two projects, partly with the aid of government regeneration schemes, but at the end of the day the results are similar in principle – that 'the have-nots' are those who in fact pay the price of very ambitious 'politics of planning and development' that are driven by capitalist interests in London and nationalistic interests in Jerusalem, as seen in the next section.

Jerusalem: urban planning and global/local nationalism – whose power, whose knowledge?

I have chosen to start this section with the short history of Elizabeth's house because it reflects more than anything else the complicated urban history of

Jerusalem. The history of her house seems to contain the major elements that influence and conduct the city's current everyday life:

> '*I live in a building that was owned by the Palestinians. The building is full of holes from the bullets of the 1948 war; it used to be the house of a Palestinian judge. In the War of Independence (the 1948 war) it served as a base of the Palmach (the Jewish Army before 1948). I don't know if the Palestinian owner escaped or was evicted; he lived on one floor and his daughter on another floor. My neighbour was in the Palmach and after the war of 1948 they told him to go to the Katamon [the name of the Palestinian neighbourhood] and 'to choose' a house. So he had chosen one floor of this building. My neighbours downstairs also moved in 1948 and then another family, a religious family, got in and 25 years ago they divided this floor; they tossed a coin and decided who would take which side and then registered it in the Taabu (land and house registry). I bought my flat from the religious family. He then said to me: you give me money for the house, which I didn't pay for . . . How do I feel about it? I used to think that I would never live in a house that was owned by Palestinians but now I think that in a situation of peace if the Palestinian owners would like to return to live in the house I could get compensation from the government and leave. I would have no problem in leaving the house; I don't feel I belong, and I would have no problem in giving it back to its old owners. It will be so great if we could reach this stage (in the peace process).'* (Elizabeth, 50s, British – Israeli–Jewish, Jerusalem, 16 March 2000)

Elizabeth's narrative reflects the tragically shared history of both Palestinian homeowners in West Jewish Jerusalem and Jewish homeowners in the Palestinian Old City of Jerusalem. These 'replaced spaces' are becoming one of the major disputes on the future of the city.

Jerusalem, the capital and the largest city in Israel, consisted in the year 2000 of 657,500 inhabitants, 11 per cent of Israel's total population. The city extends over 1.26 million hectares. It consists of the West – predominantly Jewish, and the East – predominantly Palestinian. East Jerusalem has been occupied and annexed since the 1967 war under Israeli law, bringing it under its sovereignty and making the whole and unified Jerusalem the capital of Israel. The Palestinians reject the city's unification by force and see East Jerusalem only as their capital, although they use West Jerusalem services.

These current conflicts for dominance of the city can be seen as another chain in the city's long history of battles and conflicts because of its holiness and contestation from time immemorial. Jerusalem is a thrice holy city; being claimed by Jews, Christians and Muslims (see Fig. 4.6). The Jews' attachment to Jerusalem dates back to biblical times when King David conquered the city towards the end of the eleventh century BC. For Christians, Jerusalem's holiness derives from Jesus having lived in the area and his crucifixion there. For Muslims, Jerusalem's primary religious significance comes from Muhammad's miraculous voyage from Mecca to Jerusalem and from there to heaven. This holiness has been a driving force throughout the city's history.

Figure 4.6 Holy spaces in the Old City of Jerusalem – Photographer: Jonathan Mizrachi

From the twentieth century onwards the religious issue regarding Jerusalem gradually became overshadowed by the emerging struggle between two national groups, Jews and Palestinian Arabs. Religious beliefs and symbols have often been intertwined with the conflicting aspirations of Zionism and Arab nationalism, with Jerusalem becoming the major focus of contestation (Romann and Weingrod 1991). Its contested status is expressed in the fact that even its status as the capital of Israel is not fully internationally recognised. Most countries do adhere formally to decisions of the United Nations from the 1940s that Jerusalem should have international status and recognise only that cease-fire lines of 1948 and 1967 gave control over Jerusalem to Israel and thus most countries' embassies (including the United States, United Kingdom, France) are not in Jerusalem but in Tel Aviv (Sharkansky 1996). Perhaps because of that some Israeli governments have felt that its special status and religious and political symbolism necessitates the creation of a government Ministry for Jerusalem. Politicians wanting to assert their concern for the leading city of Israel often touted this idea. The first ministry was created in 1991 but because of the increasing criticism it was abolished in 1992 and never re-established. However, this shows the importance and also perhaps the lack of security that governments of Israel attribute and feel towards the issue of Jerusalem.

Because of its uncertain status as the capital of Israel in the eyes of international law, the 'politics of planning and development' in the city have been targeted to maintain a Jewish majority in the city in what is termed 'the battle over demography'. The demographic balance in the year 2000 shows the success of these policies: 68 per cent of the city's population are Jews, 32 per cent are Palestinians. The ratio between the two populations changes in different parts of the city but it is always a Jewish majority, which dominates each area, even East Jerusalem. For example, in the city centre area (East and West) (See Fig. 4.7) there are 76 per cent Jews and 24 per cent Palestinians. The total population in this area is 117,000 inhabitants (these figures include the Old City, and Jewish neighbourhoods in West Jerusalem within this radius). In a distance of up to 4 km from the city centre the population consists of 64 per cent Jews and 36 per cent Palestinians. The total number of people living in this area is 522,000, which is 79 per cent of the total population. In the third tier of 4 km and more from the city centre there are 157,000 inhabitants, 84 per cent Jews and 16 per cent Palestinians (The Municipality of Jerusalem 2000).

This 'balance' of majority Jews is kept in two ways: first, a massive expansion of the newly Jewish neighbourhoods or settlements (a terminology that changes according to the political affiliation of the speakers), which were built after 1967 on Palestinian lands surrounding Jerusalem as part of Israeli government policies to Judaise Jerusalem. Second, in policies of discrimination towards the Palestinians citizens of Jerusalem, which are expressed in two ways: the prevention of their neighbourhoods' and villages' expansion and the discrimination in municipal budget allocated to municipal maintenance in East Jerusalem. These two points will be elaborated on later in the chapter.

East–West divisions have their distinct visual expressions too: the modernised style of West Jerusalem and the Muslim style of East Jerusalem. The distinction between the two sides of the city is reflected not only in architecture but also in the different landscapes, the different development style, the different lifestyle and politics and the sharp gaps between the population's standard of living and municipal budgets. But nationality is not the only division that affects cityscape; it is also the division on a religious and an ethnic base among Jewish neighbourhoods, especially between secular and religious neighbourhoods. The city is a collection of ghettoes, mostly self-imposed. Palestinians and Jews, as well as secular and ultra-orthodox Jews, prefer their own neighbourhoods, schools, shopping areas and newspapers (Sharkansky 1996). And the 'politics of planning and development' of the municipality allow such gettoisation to take place to such an extent that some of the ultra-orthodox neighbourhoods are becoming forbidden spaces for secular people to reside in and secular women to use (see Chapter 8) with no interference from the municipality.

This social and cultural mosaic and especially the increasing dominance of the ultra-orthodox Jews in the city's management is perhaps one of the motivations of the Jewish secular population to leave Jerusalem. The negative out-migration from Jerusalem has increased from some 2,846 people who left the city in 1990 to some 8,000 who left in 1999 (Statistical Yearbook of Jerusalem 2000). There are no indications in the official statistics to the religious affiliations of those who left the city but it is well known that most of them

Figure 4.7   Map of Jerusalem

are secular Jews. Perhaps this is one of the reasons why Jerusalem's population is poorer, suffers higher unemployment and lives in a higher housing density than the average Israeli.[7]

Indeed the economies of the city do not take a leading part in current globalisation movements. Moreover, as already mentioned above it is considered a 'poor city', which depends on donations (sent from overseas) and budget allocations of the national government concerned to maintain Israel's hold over the city. The workforce is heavily engaged in the provision of government and other public services. Bureaucrats and politicians are more prominent than private capital or business firms (Sharkansky 1996). Public and private investments work closely in the development of the city. This was especially true in the period of Teddy Kollek, the former Mayor of Jerusalem (1966–93). Three quasi-government organisations were prominent: the Jerusalem Foundation, the Jerusalem Development Authority and the Moriah Company. The Jerusalem Foundation was in fact a private foundation based on heavy fund raising, mainly from Jews in the Diaspora. At the end of Kollek's regime it had supported a thousand projects, from neighbourhood playgrounds to building museums, sports fields, programmes for dance, the arts and drama, mostly for Jews, but it also invested money in bringing together Palestinians and Jews and in refurbishing churches and mosques side by side with synagogues. Its financial contribution to the city amounted to some $245 million from 1966 to 1991, which is some 10–15 per cent of the municipality budget (Sharkansky 1996). Kollek used to extract money from the municipality and the government to match the non-governmental money that had been raised by the foundation.

The Christian and the Muslim communities in Jerusalem have their own well-established international financial support as well. The major Christian churches built their own cathedrals, monasteries, hospices, hospitals and schools in recent times. The Muslim religious trust (*waqf*) has extensive landholding in West Jerusalem that supports the Muslims' various activities in the city. Funds to Muslim institutions come from the Palestinian Authority and various Arab countries (Sharkansky 1996).

We can see that the major forces, which shape Jerusalem's cityscapes, reflect religious and national interests, and find their expressions in the ways and means in which the 'politics of planning and development' in the city are articulated.

The planning system in Israel and Jerusalem – whose power, whose knowledge?

The planning system in Israel is based on the 1965 Planning and Construction Act, which reflects a 'modernised' version of the 1936 British Mandate Planning Ordinance. The Act determines a top-down, formal hierarchy of statutory master plans under which local construction plans should result from the local comprehensive plan. The latter derives from the regional master plan that stems from the national level. In reality comprehensive planning appears to be rather static and out of date and in fact comprehensive master plans are chasing the actual development instead of leading it. Research carried out by the governmental planning administration in 1997 discovered that the average preparation time of regional and local master plans was about nine years. The same research revealed that 33 per cent of local statutory master plans which existed at the time of the research were 17 years old, or more. Only 28 per cent were 5 years or less. Moreover, 23 local master plans were in preparation at that time, 16 of them started the preparation processes 15 (!) or more years ago. This data tells us that in many cases the regional and local comprehensive master plan in Israel is practically non-existent and that urban regeneration and redevelopment just happens, not necessarily with connections to plans.

During the 1990s the formal planning administration initiated several amendments in order to flex its muscles. Among those initiatives are the Housing Construction Committees Act (VALAL) established in 1991 as an amendment to the Planning and Construction Act. These were officially established to cut short the long procedures of plans' approvals and in order to prepare the ground for the 1 million immigrants from the former USSR that were expected to arrive at the beginning of the 1990s. But in fact these procedures also cut short the already limited options to object to plans. In the end plans were approved and implemented with not enough attention paid to environmental issues, social impact, sufficient infrastructure and public spaces.

One of the important amendments in the 1965 Planning and Construction Act was the 43rd amendment, which was completed during 1996. This amendment provides those who claim ownership of the area under planning consideration, or those who have an interest in the area under consideration, to submit their own plan to the planning authorities, which in their turn have to discuss the plan. This amendment provides a legal opportunity for local initiatives in planning. Several groups of residents in Jerusalem exploited this opportunity to submit their own plan to the authorities. These procedures are reflected in people's narratives, especially as related to the notion of 'commitment' in Jerusalem.

Another change to the formal planning approaches in Israel can be seen in the 'new planning language' offered by the 35th national plan looking at 'planning tissues' rather than planning zones.[8] The 35th national master plan has not yet been approved by the government because of the objections raised from various interest groups, who make the process of approval longer than expected. However, in spite of all these changes, the local construction plan, traditionally located at the bottom of the planning tree, is still considered the strongest and the most relevant of all the hierarchical plans because this is the only plan that materialised as compared to those further up the hierarchies, which are usually changing constantly. The bottom line of this situation is the crisis of the top-down planning system. Concrete initiatives that stem from

both developers and public administrators are shaped into town or constructions plans and find their ways to be realized with the help of the local administration, almost without interference from the formal planning system. One
emphasis of these tendencies is reflected in the changing policies of the Israeli
Land Authority, the most prominent actor in shaping the built environment,
from a rigid and conservative policy of preserving agricultural state land to an
economic and financial policy that perceives land for development as an asset
of maximised revenues as a target of development.

The official city planning document is the City Outline Plan, which indicates
the future allocation of land use in the municipal borders according to the
expected and projected demand for housing, open spaces, areas for occupation,
infrastructure, etc. These entire land uses are expressed in both a map with
areas allocated to the various land uses and 'a text' that provides the technical
explanations of the map. It is usually a standard text with specific references
to the specific plan. The text is usually professional–technical, consisting of
terminologies that are unfamiliar to the general public. The plan is prepared by
independent planners (usually architects), who work outside the planning
apparatus of the municipality or the government ministry but usually carry out
most of the planning projects for them.

The planning process of the City Outline Plan is carried out by a team of
professionals attended by a steering committee, which consists of representatives of various government offices. Their duty is to make sure that the interests they represent are expressed in the plan. The plan is then discussed in
the local and district committees, in which more alterations can take place.
Before the plan is approved it is deposited for the public to allow submissions
or objections. This is a very problematic process, as the public could learn
about planning projects and the right to object to them only from special ads
published in newspapers.[9] This is in fact the only form of participation that
exists in the Planning and Building Law in Israel, which is another indicator of
its centralistic character.

The planning system in Jerusalem

The city's planning system is complex and different from other Israeli cities
primarily because it has to deal with a large number of 'actors', local and
global. Land tenants in Jerusalem include Palestinian and Jewish private
owners, churches and *waqf* owners and the state. This structure of landownership, together with the fact that Jerusalem serves as a focus of global, religious
and local–national interests makes city planning and management a complex
procedure. In this respect Jerusalem is a unique case, with such a large number
of powers, global and local, involved in the city planning and management (Bollens 2000), and because of its nationalistic symbols and ideological
emotions regarding space and territory, its development concerns a deep
involvement of international, national and the local planning authorities.

Similar to the Israeli planning system, the Jerusalem planning system consists of two main channels. The first is the formal, statutory system, a part of the hierarchical administration by law system, led by the Ministry of Interior. The second is the local initiative developmental system, which is more dynamic, pragmatic and active and includes other 'actors' such as the Ministry of Housing, the Israeli Land Authority and private entrepreneurs. These initiatives have to be compatible with the official statutory plan of the city and to undergo the official administrative procedures for their approval. However, developers often try and sometimes succeed in using their connections with national or local elected officials to overturn the planners' objections to their own initiatives. Sometimes they win their case on the claim that a project will provide substantial tax revenues to the municipal treasury. Sometimes they assert that an international hotel chain will add its prestige to Israel and Jerusalem. The Jerusalem Plaza Hotel for example was built by a well-connected group of foreign investors on the site that had been zoned to remain part of the only sizeable public park close to the central business district (Sharkansky 1996). At the city level, the main actors are the Local Planning and Building Committee, which is subordinated to the Jerusalem District Planning and Building Committee. This is the planning body where most of the national government decisions regarding Jerusalem can be and are taken without the consultation of the municipality (Bollens 2000).

Although the city has been officially 'united' since 1967, it has neither a comprehensive outline plan for the 'unified city' nor a 'master plan' for the entire city, although such a plan is under construction. The first plan to address the unified city was the 1968 Master Plan for Jerusalem, which has never been approved as such, and no subsequent town planning scheme has been developed from it (Bollens 2000). Subsequently a comprehensive outline plan for Jerusalem was prepared in 1975, approved by the local committee in 1977 but rejected by the District Committee (B'tselem 1997). The Outline Plan still in use today until a new one is approved is the 1959 plan, which was set up 10 years before the 'unification' of the city. Hopefully this new outline plan for the city will bridge the gap between the massive planning and development of Jewish projects in East Jerusalem and the necessity for a broader perspective of the city's development, both for Jews and Palestinians.

Decentralisation and planning in jerusalem

The municipality functions at the local level by means of centralistic bureaucracies similar to the way that the central government functions in the state. The municipal system is central to the extent that Jerusalem's municipal employees can make arbitrary decisions that are fateful for individual citizens, without explaining their reasons or providing access to the information that is used in making the decisions (Sharkansky 1996).

Side by side with these strong centralistic tendencies are some local initiatives, which force the system to decentralise. One of these is the initiative carried out by several of Jerusalem's neighbourhoods to establish associations of residents; some function in Palestinian neighbourhoods. These associations are funded by donors outside Israel, such as American Jewish donors for Jewish associations and Saudi Arabia, Jordanian and PLO donors for Palestinian neighbourhoods. The Palestinian associations serve as a vehicle for the residents to express their concerns in a setting where they refuse to vote or otherwise take part in activities of the Israeli state or municipality. In 1991 there were 13 neighbourhood associations with 100 employees, but today most of them have ceased functioning.

Another official effort to decentralise government in Jerusalem happened when Teddy Kollek tried to divide all of Jerusalem into the London borough system with the hope that it would provide each of them with a measure of self-government. This, he thought, would be a pragmatic way to deal with the city's diversity while maintaining the city united under Israeli control. The idea was opposed by orthodox Jewish politicians, who had seen it as a way to evade Sabbath regulations in secular neighbourhoods, and by right-wing politicians, who had seen it as the first step in the redivision of Jerusalem into Jewish and Palestinian areas. In 1992 the idea re-emerged but was again rejected for the same reasons.

Another programme that was meant to improve decentralisation in local government in Israel was the neighbourhood renewal project created by the late Prime Minister Begin in 1978. This project aimed to serve as a focus for fund raising among overseas Jewish communities, to alleviate the social and economic problems of poor urban neighbourhoods and small towns, and to involve site residents and overseas donors in project planning and management. The attitudes towards this project were mixed, both on the part of residents and the municipality. Several Jerusalem neighbourhoods have benefited from the project in terms of physical improvements in run-down housing as well as social programmes, but there have been also frustrated confrontations between local residents, representatives of overseas donors and Israeli bureaucrats over the components and the priorities that the project set up. During the mid-1980s there emerged another initiative, this time outside the establishment, to set up neighbourhood government, which would represent the residents (Hasson *et al.* 1995). The aim was to promote urban democracy and increase decentralisation and self-management in the neighborhoods. There were 14 such neighbourhood governments in Jerusalem in 1995.

Additionally, there is a large tier of NGO activities functioning in the city; some work in Jerusalem and others work on a national scale but their headquarters are in Jerusalem. They are financed by special donors and work on various issues usually related to human rights, citizen action and participation.

Nationalistic power and planning as control – the 'politics of planning and development' in Jerusalem

As already mentioned, one of the main targets of the politics of planning and development in Jerusalem is to maintain the Jewish majority in the city. The two main channels to realise this goal are a vast Jewish development and expansion and a lack of Palestinian development and improvement. The result is that although Israeli governments declare Jerusalem united, the city is managed one-sidedly (Yiftachel and Yacobi 2002). This means that urban government, urban economies and services are targeted towards the Jewish inhabitants in spite of the fact that the Palestinians constitute more than a quarter of its population.

Practically speaking, most economic and planning efforts are targeted at expanding and modernising Jewish areas as part of the policy of Judaisation of large parts of East Jerusalem and its surrounding hills. The means to achieve this goal are many: expropriating Palestinian lands, construction of Jewish neighbourhoods or settlements on these lands, restrictions on Palestinian building and land use through the adoption of planning policies, residency regulations and other measures, determining restricted housing capacity for the Palestinian population while encouraging the Jewish population by means of financial subsidies to move to Jerusalem, declaring 'open landscapes areas' near existing Palestinian villages thus preventing their natural expansion, and converting houses already built in these areas into illegals and at the same time building Jewish neighbourhoods (such as Pisgat Ze'ev, see Figure 4.7) on areas previously declared green areas. Indeed, these policies were successful. In 1993, for example, the Israeli government announced the Jewish population in East Jerusalem to be 155,000, surpassing the Palestinian population of 150,000 in the same area (Malki 2000).

Another expression of these policies is that Jewish religious people started to reside in Palestinian neighbourhoods and within the Muslim quarter in the Old City. The Ministry of Housing in particular carried out this policy over the years. These policies were sometimes initiated and implemented against municipality policy, especially during the leadership of the former mayor Teddy Kollek. Since his successor Ehud Olmart was elected in 1993 these policies have become more dominant, with the government and the municipality working hand by hand to implement various policies concerning the Judaisation of the city. Between 1967 and 1995, 76,151 housing units were built in Jerusalem, of which 64,867 (88 per cent) were for Jewish residents; 59 per cent of these were built in new Jewish neighbourhoods in East Jerusalem (B'tselem 1997).

Another expression of the policies of the Judaisation of Jerusalem is the determination of its municipal boundaries. In spite of its being a planning issue, the boundary drawing of Jerusalem was not made by planners but rather by Israeli Defence Force (IDF) officers with the purpose of including the maximum territory possible with minimum Palestinian population (Cheshin *et al.* 2000).

As the primary purpose of the boundary drawing was political, that is, to include maximum land and exclude the maximum number of Palestinians (Bollens 2000) with no consideration given to planning perspective, the city's boundaries exclude areas that functionally belong to the city, thus creating 'an artificial construct' (Faludi 1997) and legitimising any development as municipal and not in the 'occupied territories'. Thus Jordanian (pre-1967 rule) East Jerusalem consists of only 6 sq km but Israel annexed an area of almost 70 sq km to the municipal boundaries of West Jerusalem, of which 64 sq km consists of 28 villages in the West Bank surrounding the city, obviously for political purposes (B'tselem 1997). At the same time Israel has deliberately left out newly created East Jerusalem's neighbourhoods such as Abu-Dis that the Palestinians see as an integral part of the city. This is in spite of the fact that East Jerusalem has for decades been the political and spiritual centre of the Palestinian national movement as well as the geographical and economic link between the Northern and Southern parts of the West Bank and between the West Bank and the Gaza Strip (Ma'oz 2000). The 1993 closure was another trigger in disconnecting East Jerusalem from its hinterland. Bus companies, which operated regularly from East Jerusalem to the cities in the West Bank, were forced to reduce their trip frequencies because of the closure and especially since the rise of the second Intifada in October 2000 (Malki 2000).

The Palestinians in the metropolitan region of Jerusalem are given residency rights but not Israeli citizenship. As residents they carry the 'blue' identity card like any other Israeli citizen. Their blue ID card allows them free movement within Israel, which is forbidden for the Palestinians who live in the West Bank and Gaza who do not carry blue ID cards. Moreover, it allows them to work within Israel whereas the Palestinians living in the West Bank and Gaza Strip need to have special permissions to work in Israel, a situation that has become more and more difficult as the latest political conflict accelerates. But this blue ID card also serves as a mechanism which separates them from those living in adjoining localities that remained in the 'occupied territories' of the West Bank and Gaza Strip with no residency or movement rights in the city or in the country at large. This situation has a double effect: the Palestinian residents of Jerusalem are excluded from the city's forums of decision making due to their refusal to accept the imposition of Israeli law. At the same time, the Palestinians living in the adjoining localities cannot use the city as an urban centre because they have no movement rights. As a result East Jerusalem ceased to function as the urban metropolitan centre for its Palestinian hinterland and its role as the capital of the West Bank ceased.

These policies of control reflected in both planning and development and in the restrictions of freedom of movement means the Palestinian situation is an absurdity. In terms of their own planning rights, it is impossible for them to build their houses within the legal frameworks of their localities because the 'blue line', that is the boundaries of their neighbourhoods and villages, was determined so that there is no room for new houses to be built. The demographically contested manipulation is becoming an arena of municipal regulation and building permit issues. Between 1968 and 1974 only 58 permits were issued. In recent years about 150 permits per year were issued. The total number of permits that were issued between 1967 and 1999 was 2,950, when the Palestinian population grew from 68,000 to 180,000, which means that out of the total 19,650 Palestinian homes built in East Jerusalem after the 1967 war, 16,700 do not have permits and are therefore considered illegal and face demolition threats and heavy financial fines (Amnesty International 1999).

Practically it means that nearly one-half of the Palestinian population of East Jerusalem live under threat of having their houses demolished. Thus the current Israeli policy excludes the majority of Palestinians in Jerusalem by making their position illegal. In reality, however, only a small number of demolitions were carried out: 284 houses out of 12,000, or 2.3 percent. Each demolition is widely covered by the media in Israel and abroad and is followed by protests by human rights organisations, both Israeli and Palestinian. At the same time there are many additions to private homes in the Jewish sector without the formal approvals that are necessary (Sharkansky 1996). These policies have different labelling in the literature. Faludi (1997) calls them the 'political doctrine of planning' and Bollens (2000) defines this situation as the 'partisan approach to urban planning'. This means focusing on planning and development for one group of residents, the Jewish in this case, while discriminating against Palestinian residents according to Israeli law. This labelling of illegality provides Israeli authorities with the right under their laws to demolish buildings for lack of building permits. Bollens (2000) calls it the 'municipalisation' of what in reality are decisions with international connotations. Political planning serves here as a central tool in the creation of discriminatory situations in Jerusalem through those exclusionary policies and projects. This situation resulted not only in deep discrimination and frustration among the Palestinians and Jews living in the city but also in increasingly international criticism and even condemnation.

In spite of these discriminatory policies the Palestinians remain in the city and stay to live on their land. This is called *zumud* – staying on the land, resistance, ongoing presence in the city and around it and building as a major part of it. It is not an independent policy with explicit goals as some of the Israelis see it. Rather it is spontaneous, not coordinated or planned in advance by the national or local Palestinian leadership (Klein 2001). The story of Isawiye, a Palestinian village located at the outskirts of Jerusalem, illustrates the 'national political planning narrative', which dictates the everyday life of the Palestinians in Jerusalem.

Nationalist power and control in Isawiye village planning narrative

The planning history of Isawiye illustrates the tragic situation of the Palestinians in Jerusalem. Isawiye is located east of the French Hill – a Jewish neighbourhood built on their expropriated land (see Figure 4.7). As Jewish development expands in the area their land is more and more in demand by the municipality. Before 1967 the residents of the village owned some 10,000 *dunams*[10] but some of their land has been appropriated by the Jordanians so that after 1967 when the area became part of the Israeli occupation they possessed only 2,700 *dunams*. The location of the village in proximity to the renewed Hebrew University in Mount Scopus made it attractive for developing Jewish neighbourhoods, and thus some 800 *dunams* of land were expropriated by the Israeli authorities to build the French Hill Jewish neighbourhood.

But this was not enough, since the municipality was still interested in expropriating more land, as Isawiye's lands had become part of the city's outskirts. For that purpose the municipality prepared an outline plan for Isawiye, which includes only between 680 and 700 *dunams* as the village's 'legal' area out of the 1,900 *dunams* they possessed, while the rest of the village lands (some 1,200 *dunams*) were intended to be expropriated for Jewish development in the future (Kaminker, 31 October 2001). Moreover, the municipality's planners turned a large area of the village into a 'green zone' although it is a residential area at present. Thus the houses in the area are becoming illegal although they were built many years ago. The planners allocated at present only 270 *dunams* for housing in an area that is already built (mainly the village centre) and which is the most densely populated area in the village. The residents of Isawiye want the municipality to plan their village on the whole 1,900 *dunams* they own and not on the 270 *dunams* that the authorities planned. This is mainly because the area the municipality planned for them is already heavily populated and therefore the plan does not meet their housing needs. The Director of the Department of Planning Policy at the Jerusalem Municipality explains that the principle of green belt area is actually inherited from the British planning system. The green belt area principle appears in a plan prepared for East Jerusalem in 1919 and in another plan for Jerusalem prepared in 1944. The idea of the green belt area was to preserve the unique visual aspects of the Old City by preventing any construction and development in the areas declared as green belt preservation areas (interview, 2 February 2002). As it happened these areas fall within Isawiye lands and thus are officially used as the reason to prevent housing construction on most of its lands. This is an illustration of how old colonial perceptions sometimes find their way to become tools and mechanisms of control over minority groups, reflecting perhaps 'new colonial expressions'.

The outline plan of Isawiye, which has been prepared without consulting the residents, was published in 1992. The residents objected. Their leader says: '*These are stupid plans, political plans. The plans are not meant to serve the population needs.*' (14 November 2001). The level of housing density is becoming intolerable and many houses are becoming illegal (see Figures 4.8 and 4.9). The residents organised themselves with the help of a freelance urban planner to prepare an alternative outline plan to the village, possible according to the new 43rd amendment in the 1965 Planning and Construction Law (see

Figure 4.8   Isawiye's living density and the Hebrew University Tower

previous section). Using this amendment the residents and the freelance planner have tried to get the funds to prepare such a plan. This process has been
halted since the beginning of the second Intifada in October 2000.

The hardship the residents of Isawiye face is expressed in what their leader
says:

> 'We're lacking in the village every aspect of infrastructure: roads, schools, street
> lights, sidewalks, public parks, community centre, we have no postal services, no
> water supply . . . I asked the municipality so many times to do something about
> it . . . we have the village's main road that the municipality started constructing
> in 1997 and haven't finished today. Loads of holes in the roads . . . there is no
> appropriate drainage system and in winter there are floods. When we talk to the
> municipality they say they don't have enough budget so I ask them how come you
> have enough budget for the Jewish neighbourhood – "the French Hill"? They say:
> Don't talk politics! So I say: those who demand their rights talk politics? We pay
> municipal taxes (arnona), as much as the Jewish neighbourhoods pay and even
> more. Why can't we get the same level of services?' (Interview, 7 November 2001)

The village leader clearly expresses the bitterness and frustration that the
Palestinian residents of Jerusalem and the outskirts suffer. On the one hand
they feel part of Jerusalem and therefore they pay taxes to the municipality
(*arnona*). On the other hand they feel discriminated against because they are
not given the services they are entitled to by their tax payments. Their sense of

Figure 4.9 One of the narrowest alleys in the village with illegal construction

belonging is complicated; they do feel residents of Jerusalem but not residents of the Israeli-occupied Jerusalem. It is a vicious cycle. They do not vote in the municipality elections as an act of protest against the Israeli occupation. But because of that they do not have their representatives in the municipality councils to watch out for their needs and rights.

When I met the people from Isawiye (November 2001) they were in the middle of a crisis. The police and the army blocked one of the roads leading from their village to the centre of Jerusalem. The reason for this blockage was stone throwing from the village to the roads in the area as part of the Intifada. As in previous cases, while the criminals are individuals the punishment is collective. The residents of Isawiye were furious and wanted to protest but their leader made efforts to calm the anger down and to sort out the problems peacefully. When I met them they were about to meet the chief of police and an IDF officer in order to sort out the situation peacefully. When we met next, in the following week, they had just came back from another meeting with the police where they had reached an agreement that the blocked road would be opened and the leaders would make efforts against stone throwing from the village area. This event is only one example of the hardship that the Palestinian population undergo in such troubled times of the Intifada, and their leaders are trapped between those youths who choose to fight against the Israeli occupation (using various means, from stone throwing to suicidal bombing) and the strong reaction of the Israeli authorities towards such acts. I was hesitant to visit the village, a place I would easily have gone only a year before. It was only when the village leader assured me that it was safe to walk around in his company that I was willing to go. Identity issues and politics have become a major component in carrying out this research in my own residential city.

To sum up this section, in Jerusalem 'political planning' has its expressions mainly in favour of one nationality over the other, using professional tools and planning mechanisms to control and oppress the Palestinian population in the city. These policies may have been successful in terms of maintaining the Jewish–Palestinian ratio in the city but they have proved a failure in that they have created strong feelings of resistance, frustration and hate and thus pushed back the chances of establishing a civil and just society in the holy city.

Summary and conclusions

London and Jerusalem – two cities so different, so contrasting and still perhaps surprisingly similar in the way 'the politics of planning and development' function so that they reflect power relations and control. In both cases power relations are the core issue in understanding the 'planning games' that shape and reshape cityscapes. Several issues have been highlighted in the chapter:

- *In London 'the politics of planning and development' involve the capitalist powers*, which influence and sometimes even dictate the organisation of city spaces and badly affect local communities such as the Bangladeshi in East London. This battle involves the dominance of the developers, government agencies and local communities.

- *In Jerusalem 'the politics of planning' express nationalistic goals* for the Judaisation of the city using professional tools and planning mechanisms to control and oppress the Palestinian population in the city. This battle involves the ideological/nationalist hegemony and dominance of the Jewish establishment and the Palestinian local communities.

- In spite of the enormous differences between the two cities the analysis of these two cases show *how planning uses the same old tools for control*: land expropriation, zoning principles, planning gains, etc., to win this battle. In both cities planning has become narrowly functional-oriented action rather than comprehensive. This situation is similar to Harvey's (1987) view of planning as coordination of the interests of business while the planners play the role of facilitators of specific projects rather than extending this comprehensively to the relationships of projects to each other and to the housing and transit systems.

- *Jerusalem's triple holiness made it an important 'global' city from a religious point of view.* Churches have land assets in the city, which make them one of the active actors of planning games in the city. There are private–public deals between the government of Israel and the various churches that function in Jerusalem as real estate owners.

- *The struggle for identity and the place of 'the Other' in the two cities takes place in the context and arena of urban planning and governance.* In the United Kingdom urban planning and governance has not yet incorporated urban policy responses or recognition of difference at the national level. This was true especially during the Thatcher era when the driving force of urban governance was development with a single-minded commitment to private capital (Sandercock 1998a). Even what Sandercock perceives as indications of opportunities for the Bengali immigrants' involvement in the development processes in the Bishopsgate Goodsyard project and small steps towards cosmopolis seems not to be fulfilled at the moment.

- *In Jerusalem the lack of recognition of difference reflects both at the national and municipal level.* It is a national–territorial ideology, which uses urban planning and governance to reduce spaces of citizenship for the Palestinians by cutting notions of comfort in their everyday life. Politics of difference are by and large far from realisation as long as the government of Israel and the Jerusalem municipality fight for preserving the sole Jewish identity of Jerusalem's territory against the Palestinian identity of its residents.

- *Minorities in Jerusalem and London are facing the same struggle for recognition and the rights for identity and place.* In Jerusalem the struggle is much more fierce and touches basic human rights issues. In London the struggle is about a sense of belonging and sense of home that the Bangladeshi wish to create, an effort that sometimes does not coincide with developers' interests and local councils' goals.

- *Capitalist powers appear to leave more room for manoeuvre than nationalist powers but in practice the two components have the same power* to exclude, discriminate and oppress the 'Other'. If we return to one of the questions posed in the introduction, 'are all cities alike?' it seems as if some of the major forces that motivate cities' development and expansion are indeed alike.

Notes

1 Most of the figures and information that appear in this section are based on the Internet site of the new GLA (www.london.gov.uk/mayor/case_for_london/ index/htm).

2 This section is largely based on Fainstein, S. (1994) and interviews with Kay Jordan, Director of Spitalfields Small Business Associations; Micheal Parks, a Community Planner involved in Bishopsgate Goodsyard Community Plan; Nisar Ahmed – Urban Regeneration Expert, London; Michael Safier, Development Planning Unit, University College, London, Professor Derek Diamond – Professor Emeritus of Urban Planning, London School of Economics and Political Science, London.

3 This section is largely based on interviews with Kay Jordan, Director of Spitalfields Small Business Associations; Micheal Parks, a Community Planner involved in Bishopsgate Goodsyard Community Plan; Nisar Ahmed, urban regeneration expert, London.

4 The information appearing in this section is based on an interview with Nisar Ahmed (28 August 2001).

5 This group, consisting mostly of English white middle-class people, managed in several years to purchase most of the houses and to create a 'self-organising' gentrification of the area.

6 Grand Metropolitan PLC is a large conglomerate with interests in land development as well as food and leisure activities.

7 Consequently it is Tel Aviv, the second largest city in Israel, that takes the leading role in the globalised economy. Tel Aviv was declared one of the ten hi-tech centres in the world (*Newsweek* 1998) with a high level of Internet users (some 360,000, with an annual growth rate of 30 per cent (*Yediot Hachronot* newpaper 1998). The difference between the two cities is reflected in recent figures on economic development. For example, during 1990–98, 57 per cent of the construction in Tel Aviv was housing construction and 27 per cent offices and businesses. In Jerusalem during the same period 65 per cent of construction projects were housing and only 9 per cent offices and business. The differences between the two cities are reflected in their budget resources. Of the municipal taxes in Jerusalem 40 per cent come from dwellings while in Tel Aviv only 15 per cent are covered from dwellings while 50 per cent are covered by business taxes.

8 Planning tissues are areas of a mixture of activities, as opposed to zoning areas, which consist of one major land use to an area.

9 These ads usually appear on the back pages in very small letters with 'planning' information that is not known to the majority of the public, for example: the location of the area where the planning takes place is indicated in terms of region/plot number rather then simply indicating the street name and number. Recently this latter information has become part of the text presented in the advertisement but the ad is still not the main means by which people learn about developments in their area.

10 One hectare is equivalent to ten *dunam*.

The Local Embodied Knowledge of Comfort, Belonging and Commitment in the Global City and the Holy City

Are all cities alike? Do people experience the same everyday life in such contrasting cities as the global, world city of London, and the holy city of Jerusalem? The debate around the commonalities and differences of everyday experiences and local embodied knowledge of people living in cities is highlighted in Part II.

It deconstructs the three central elements consisting of local embodied knowledge: comfort (Chapter 5), belonging (Chapter 6) and commitment (Chapter 7). It provides a general analysis of the three elements by way of comparing people's narratives in Jerusalem and London and analysing the relationships between them in order to identify the similarities and differences in people's experiences in the two cities. In Chapter 8 the three elements are analysed with regard to gender identities.

Narrative analysis as a method for analysing the meaning of comfort, belonging and commitment

In analysing people's narratives regarding comfort, belonging and commitment to the environment they live in I have chosen to use the narrative analysis method. This method seems helpful to identify people's associations, feelings, perceptions, desires and needs regarding these three elements. I have used this method to build up the framework for analysis so that these experiences become local knowledge. Similar approaches of using people's stories in planning have already been discussed in the literature dealing with planning (Forester 1989, 1999). Sandercock (2003) put a special emphasis on the importance of story in planning practice, research and teaching. Story for her: 'conveys a range of meanings, from anecdote, to exemplar, to something that is invented rather than "true", in the sense of strictly adhering to widely agreed-on facts' (p. 3). The narratives in this book are stories in the sense that they tell us without intermediacy what people experience in their daily practices.

The narrative analysis method

Narrative research is a common method in qualitative research. Narrative research is defined as referring to any study that uses or analyses narrative materials (Lieblich *et al.* 1998). This method is widely used in psychology, sociology, medicine and anthropology and, in addition to the works mentioned above, another that links the fields of architecture and linguistics, is by Markus and Cameron (2002). Their main argument is that 'the language used to speak and write about the built environment plays a significant role in shaping that environment, and our responses to it' (p. 2). Languages are after all the richest symbolic systems to which human beings have access and that they use for communication. We use our language as an aid to and expression of our thinking. These activities, argue Markus and Cameron, are part of designing and planning the built environment, together with other tools such as mathematical calculations. Indeed over the years architects and planners have developed linguistic resources for the purpose of reflecting on their thoughts and ideas. These fields became what Markus and Cameron define as *linguistic register*, which is mainly extensive technical vocabulary, known mostly to those in the professions.

As already mentioned in Chapter 2, the Foucaldian way of looking at this *linguistic register* is as a form of exclusion and is a reflection of power/knowledge relationships, usually between planners and architects and the users of their planned spaces. The studies of narrative analysis and discourse analysis derive in many forms from semiotic and structuralist approaches (Forty 2000), which are concerned not only with what things mean but also with how meaning occurs. As argued by Markus and Cameron (2002), both buildings and language are irreducibly social phenomena so that any analysis of them must locate them in the larger social world. I would add that this is true of any practice of urban planning, not only at the level of a building but also at any spatial level. Urban planning thus is social and also a political phenomenon (see Chapter 2) and thus any analysis must locate language and urban planning within both the social and political context.

Models of narrative analysis

Two distinct dimensions are dominant in the narrative analysis method: *the holistic versus the categorical approach* and *the content versus form approach*. The many variations that exist in reading and analysing text represent the combination of these approaches (Lieblich *et al.* 1998). An intersection between the approaches suggests the following four modes:

1. *The holistic-content mode* of reading uses the complete life story or narrative of an individual and focuses on the content presented by it.

2. *The holistic-form-based mode* of analysis looks at the plots or structure of complete narratives – how the narratives develop. The researcher may look for a climax or turning points in the story.

3. *The categorical-content approach* is more familiar as 'content analysis'. Categories of the studied topic are defined and separate utterances of the text are extracted, classified into categories/groups. Categories may be very narrow or broad.

4. *The categorical-form mode* of analysis focuses on discrete stylistic or linguistic characteristics of defined units of the narrative. For example, what kind of metaphor is the narrator using, or how frequent are their passive versus active utterances?

As already mentioned these distinctions are not always clear-cut in reality and usually a combination of approaches is used, as elaborated below.

Narrative analysis of comfort, belonging and commitment

Stages in the analysis

In exploring the meanings and symbols of comfort, belonging and commitment as inherent in people's narratives, several steps have been taken:

1. Arranging the narratives according to the hierarchy of space. In this way it is not a person's narrative as a whole that is under analysis but rather the related themes which were divided into the various categories of space. This method puts the 'category of space' rather than the 'identity of the participant' at the focus of the analysis. As I previously mentioned my main focus in this research is on the way spaces are constructed and symbolised, that is, *the outcomes* of people's experiences rather than *the process* of constructing them, which would be the case had I chosen to focus on people's narratives as a whole.

2. After arranging the narratives according to the relevant category of space (home, building, street, etc.) I have read each group of narratives through once to identify the main themes, which are repeated in people's narratives for each category of space. This follows the '*holistic-content mode of analysis*'.

3. I then read the narratives again using the *categorical-content* type of analysis, looking at the thematic fields and categories that each of the participants expressed when talking about each category of space.

4. I then identified in each narrative the metaphors, symbols and synonyms used for describing and analysing each category of space with the intention of identifying repetitive words, expressions, feelings, etc., using the *categorical-form mode* of analysis.

5. Finally, I read the text again to look at the narrative as a whole.

Categories used in analysing the narratives

After analysing the narratives following the above stages I identified four main themes or categories, which I will use to analyse the three concepts of comfort, belonging and commitment:

1. *The physical aspects and the social–emotional aspects of each of the metaphors, symbols and synonyms* used. The choice of using the physical–socio divisions is because of its relevance to urban planning and management.

Physical aspects of comfort, belonging and commitment are more associated with physical planning of the city, while social and emotional aspects are associated with other fields of urban planning, 'the soft edge' of urban planning, that is, social work, community work, welfare, psychology, etc.

2. *The private and public divides.* I wanted to find out whether the well-known private–public divides are reflected in people's narratives regarding comfort, belonging and commitment. I also wanted to explore the more delicate intricacies of the two terms in the six categories of space.

3. *The positive and negative associations of the metaphors, images and symbols* used for each category of space and each aspect (physical or social).

4. *The power relations contexts that are expressed in the narratives, whether implicit or explicit.* Here I refer to the associations brought up in the people's narratives, sometimes hidden, sometimes overt.

5. *Commonalities and differences* in identifying the concepts of comfort, belonging and commitment. Commonalities and differences are examined both as related to a comparison between citizens of London and citizens of Jerusalem and a comparison between people of different identities.

This way of analysis allows in-depth understanding of the needs and especially the desires of the people living in the two cities and as such can become a basis for planning.

CHAPTER 5

On comfort

'*Comfort for me is associated with the house environment . . . the first thing that jumps into my mind is to feel that I have control of the time that it takes me to walk from home to work, to the bank . . .*' (Eitan, 30s, married, Israeli – Mizrachi–Westerner, Jerusalem, 3 February 2000)

Comfort for Eitan is associated with control of time and space – the time it takes him to move from one place to another in his close environment. Obviously, a sense of comfort is a subjective feeling. People of different identities define and express comfort in different ways. And yet one of the interesting outcomes of this project is how similar the notion of comfort can become for people of different identities who live in the two cities. The aim of the chapter is therefore to explore both the commonalities and the differences of comfort, both its subjective–personal and objective–universal connotations, and from that to draw some basic understandings of the notion: 'the right to comfort in the global city' which was established in Chapter 3. Questions such as, what does this right to comfort include? And, are its components privately based or are they connected to the wider notion of 'the right to the city' will be discussed in the chapter.

In order to analyse these questions, the chapter begins by examining the notion of comfort in two different environments, the very private – the home, and the very public – the city. The first question I asked people in the interview concerned their associations regarding the words 'home' and 'city'. The aim of this question was to find out what are the meanings of these contrasting environments and whether it is possible at all to create a sense of 'home' in the city. The chapter then moves on to discuss the meaning of comfort in the various categories of environment identified earlier.

Home and the city as contested spaces

'Home – prison! Although in my room I have all I need to "get out" – computer, Internet, video, TV cables of 50 channels . . . I have everything but this is not it.'

'City – freedom, personal freedom, atmosphere, spring.'

This is probably one of the most striking examples, which emphasises how cultural constructs and identity issues are so central in defining the three concepts. Fatma is an Arab – a Palestinian–Israeli woman in her early 40s, unmarried, and therefore having to live with her mother. This is what she told me when I asked her about the associations with the word 'home' and 'city'. In her narrative she uses a language that symbolises more than anything else her lack of comfort at home and her sense of freedom in the city. Lack of comfort is not physical, as she herself says she has everything in her room but this is not enough for her (see Part III for the analysis of her mental map of home). Lack of comfort in her case is culturally constructed. It results from norms and values that Muslim culture sets up for unmarried women, which reflect strong control and power relations at home (see Fenster 1999a). For this woman, her unmarried status makes her home a prison, a place where rules of behaviour are very strict and explicit; it is forbidden for her to live her own life, have her own house, and be separated from her family, her community, and her cultural norms.

Her home brings up negative associations, which contrast the meanings of 'city'. Her associations with the city are all positive, refreshing. It almost seems that the city becomes her 'private' or 'intimate' space where she can be herself. 'These cities', writes Elizabeth Wilson in *The Sphinx in the City* (1991), 'brought changes to the lives of women. They represented choice' (p. 125). There she refers to the new colonial cities in West Africa, but the role of cities in women's lives as providing choice seems relevant in other places as well. These contrasts repeat themselves in many of the associations with the connotation of 'city' that the interviewees made. '*City for me is London and Jerusalem*,' says Aziza, a Palestinian woman, a citizen of Israel in her 30s living in Jerusalem, '*it is a place full of love and freedom, in particular freedom, this is my place.*'

For women in certain cultures the city is associated with what is usually a synonym for home: feelings of freedom, the possibility to be as one likes, a place full of love and happiness. For those women comfort is associated with power relations, which for them are strict and strong at home – where their culture is dominant – whereas in the city their cultural norms are less restricted because they are anonymous and this anonymity is what allows them to feel what Aziza feels in Jerusalem. For these women the right to comfort is associated with basic human and citizen rights (see Chapter 3), which are not inherent in their private and public life. For them (and perhaps for each of us) a discussion on comfort is actually a discussion on personal, social, cultural boundaries. The wider the gaps between their desires, aspirations and identities and their everyday life realities the less comfortable they feel. For some women these gaps are most explicit at home and therefore home becomes like 'a prison' (This point is elaborated on in the next section.) The right to comfort in their case is therefore a matter of a 'boundary expansion', of bridging the gaps between their felt desires and aspirations, their cultural and gender identities and their everyday life possibilities.

In other narratives home is perceived as a general term that expresses feelings for a specific place or location; home could be a particular room in a house, a flat, a whole house or a city: '*Home is millions of things, it is kids, stairs, kitchen, living room*', says a woman in her mid-40s, married with four children, living on the outskirts of Jerusalem. '*Home is Edinburgh, home is my flat*', says a woman in her mid-40s, single, Scottish, living in London. Home is a physical, social and emotional space or, as hooks (1991) terms it, an intellectual space where one is free to explore, to discover, to move, to grow, where a rigid and unchanging identity is not imposed.

'*Home is my address, my little house, my English side is coming out – my house as my castle, opposite of nomadism and movement, my stable position, my base, the house is me.*' (James, 60s, single, living in London)

We can see how different and are some of the above definitions of home from those of the Muslim women expressed earlier, which demonstrates again how home is a notion associated with power relations. The more a person feels in power and control the more positive their reflections of home. I will get back to this point later in Chapter 8.

Home as an emotional space is associated with the 'self', the me: '*Home is me, the concept of home defines the boundaries between the inner world and the outer world*' (Mary, 40s, single, Jerusalem). Home is not only the physical shelter; it is in most cases an emotional space. The 'home' and 'the self' are both culturally constructed and therefore comfort at home means in many cases comfort with one's own identity. This terminology of home reflects an expanded boundary of the 'self' (Sibley 1995), it secures privacy and intimacy. Moreover, the order of the home represents internal and emotional balance. For one woman: 'the meanings of the objects she is surrounded by are signs of her ties to this larger system which she is a part of' (Sibley 1995: 10).

There seems to be some common shared knowledge towards the notion of 'home' that cross-cuts identities. No matter what the symbolism of home it is usually a positive, relaxed space, with the exception of the experiences of some Muslim women who feel 'in prison' at home. To this latter context we can add Sibley's observation (1995: 91) that: 'when the desire for a purified environment is not shared by all members of a household, the home becomes a place of conflict, and there is some evidence to suggest that this kind of conflict may lead to behaviour problems, particularly in adolescence'. For both women and children in certain cultures home is the focus of power relations and of potential suppression.

The city is usually the contrast to home, either positively – *'city is liberating whereas home is prison'* or negatively – *'city is dirt'*, *'crowds'*, *'noisy'*, *'unsafe'*, but it is always also *'magic'*, *'vibrant'*, *'dramatic'*, *'sophisticated'*, *'it is full of contrasts'*, *'full of surprises'*.

The city has been a source of inspiration for so many academic researchers, novels and artwork, each seeking to understand and analyse the essence of the city life. In one such work (Wilson 1991) two somewhat contrasting perspectives of city life are presented:

> This recurring image, of the city as a maze, as having a secret centre, contradicts that other and equally common metaphor for the city as labyrinthine and centre-less. Even if the labyrinth does have a centre, one image of the discovery of the city, or of exploring the city, is not so much finally reaching this centre, as of an endlessly circular journey, and of the retracing of the same pathways over times. (p. 3)

Wilson associates the city with mazes, labyrinthine, endlessly circular journeys and other images that represent confusion and loss but also with changes: 'for the city is in a constant process of change, and thus becomes dreamlike and magical yet also terrifying in a way a dream can be' (p. 3).

Raban (1998) describes another aspect, the intimate aspect of the city, the 'private city':

> because it is too large and formless to be held in the mind as an imaginative whole that we make irrational short cuts and simplifications. As a result, within the city at large there are: 'multitudes of contracted, superstitious cities, sequestered places with clear boundaries, rituals and customs, whose outlines often correspond with those of the tribal castes and styles. (p. 165)

Raban's 'private city' is a synonym to what we termed in Chapter 3 'the mental city', 'the symbolic city' that emerges, of personal symbols and taboos. These images and symbolism of the 'private city' are part of the 'local knowledge', knowledge of our private city that can contribute to make our cities feel like home.

For some people their 'private city' is only positive: *'City is cosmopolitanism, diversity of cultures, food, dresses, colour, and religion . . . I just like it!'* says a retired academic male living on the outskirts of London. *'City is a good thing, it is civility, something positive, vibrant, dramatic, sophisticated, the greatest engine of improvement'*, says a single man in his 50s living in London. For others their 'private city' is a specific city: *'City is London!'* says a woman in her mid-40s, single, Scottish, living in London. *'City is London'*, says a man in his late 30s, a new immigrant in Jerusalem. Interestingly enough, none of the interviewees who live in Jerusalem said: *City is Jerusalem!* Those who live in Jerusalem mention other cities and then the city they live in, whereas those who live in London associated the meanings of city mostly with London. Here perhaps we can trace the differences between living in a global city with all its positive implications, which brings London to mind as the first association with the word 'city', whereas living in Jerusalem, 'holy' for some but not for all (probably not for the people I interviewed) does not bring Jerusalem to mind as an association with the word 'city'. It seems that the connotations of the word 'city' is more of what London represents then what Jerusalem does.

Several interviewees pointed out the diversity of the city: '*City is crowds, lots of activities, constant movement, interest, excitement*' says an Israeli woman in her late 30s living in London. '*City is coffee shops, shopping, banks, art, music, culture*', says a woman in her mid-40s, married with four children, living on the outskirts of Jerusalem. The 'private city' can also be a specific location, an 'address' within the city: '*City is the street where my office is located, it is transportation, centre*', says a man in his late 40s, married with three children living in Jerusalem.

Lewis Mumford wrote in 1945 that 'social facts' should be considered as primary to the physical organisation of a city, of its industries and markets, which must be designed so that they meet the social needs of the population. Perhaps here lies the basic foundation of the understanding of the notion 'city' vis-à-vis 'home', the city is actually an aggregate of our homes, of the social needs and desires that are the base of the city's functioning. Therefore one of the challenges of city planning and management can be identified as creating a balance between the physical and the social aspects of the city's expansions so that cities become our home in the social and emotional meanings of the word, a place where we can be ourselves.

Once the understanding of the 'home–city' dichotomies has been established it is perhaps appropriate to move onto the analysis of the meanings of comfort in each category of space.

Comfort at home

In the previous section we discussed the associations that came out from the word 'home' in contrast to the associations of the word 'city'. In this section we focus the analysis on the components of comfort at home, which brings up some more concrete understandings of the notion 'home'.

> '*Comfort at home is about my books, my music. Having my things where I know where they are, where they belong . . .*' (Stuart, 30s, single, British–Ghanaian–Afro-Caribbean, London, 6 August 2002)

Like Stuart, most people associated comfort at home with positive connotations relating to the ways their 'intimate' spaces are constructed. Their connotations were usually physical:

- *a lot of* light, space, quietness, big or small rooms, suitable arrangement of the flat, an order to things, connection with outside – windows, access to garden, good location
- *personal possessions* – books, music, small souvenirs, the order of things
- *no intrusion of* noise or smell
- *items such as* flowers, colours.

But for most people comfort at home is not only physical. *Comfort at home* is also associated with social aspects and especially emotional aspects. These aspects are related mainly to people, especially family members who live at home:

> '*Comfort for me is first of all the people who are present at home, or could be present at home, then the order to things . . .*' (Abraham, 50s, married, Jewish, 29 February 2001)

The association of family members with a sense of comfort is emphasised in contrast to the 'intrusion' of strangers, or even invited guests. It seems that these 'intrusions' break the intimacy and privacy of people in their most private space – the home. Comfort at home is associated not only with the company of human beings such as family members but also with pets.

Emotionally, comfort at home is linked with positive, intimate synonyms such as feelings of privacy, aesthetics, freedom or being in control, warmth, familiarity, harmony, sanctuary, calmness, protection, intimacy, safety, a place where one can be oneself. These emotional associations remind one of Csikszentmihalyi and Rochber Holton's approach (1981, in Sibley 1995), that for some people 'home environment reflects an expanded boundary of the self, one that includes a number of past and present relationships' (p. 10). This definition of the home as an extended boundary of oneself is expressed in Saida's narrative on comfort at home. She actually talks about the gaps between her personal needs for privacy and the realities of her everyday life that intrude on these needs:

> '*Comfort at home means my own privacy, I decide what to do, I am free. Discomfort means that something disturbs this privacy, unexpected guests . . . also having other people living in the same house, each has a different opinion so you are not as free as you should be, the house has too many people and then I don't have my own privacy . . . lack of it causes discomfort.*' (Saida, 30s, single, Palestinian – Palestinian, Jerusalem, 30 December 2000)

Here we elaborate on a point made in the previous section on the connection between gendered comfort and power relations. Saida describes the meanings of comfort at home in terms of culturally constructed power relations, which reflect a wide gap between the boundaries of her personal desires and her everyday realities. As mentioned before, most single Muslim women who were interviewed still live in their parents' home because of cultural norms, which do not allow them to expand their personal boundaries. Most of them expressed feelings of discontent with the lack of privacy they experience, as Fatma put it:

'I can not do what I plan to do. For example, I want to work on my students' dissertations or to read some material or to write and I can't do most of it because I live in a house where my mom lives and she is the mom of all my brothers who come to visit her and there is such a noise that I just can't work. Also the kids want me as their aunt, they want my attention . . . It is hard to draw the boundaries in the house.' (Fatma, 41, Single, Arab – Muslim–Israeli, Jerusalem, 18 April 2000)

These two narratives present a contrasting view of the idealised view of home expressed previously by most people interviewed. It is because for both Saida and Fatma their home does not reflect themselves. It follows Massey's (1994) and hooks' (1991) analysis of the notion of home as power relations related and in these specific cases as culturally related. Massey and hooks present another challenge to the idealised romantic notion of home, similar to Sibley's analysis of home (1995). They argue that home is not one single place but consists of multiple places. It is both singular and bounded (Massey, 1994). Here notions of comfort and belonging to home intermingle. We will get back to this point in the next chapter.

The centrality of power relations in dictating comfort at home, which is so explicit in Fatma and Saida's narratives, is not so clear in other people's narratives. Probably because most of them are 'in control', that is adults, independent women, middle-class people, for whom the notion of control at home is dealt with from 'the controller' side rather than the controlled. This perception is expressed in Colomina's (1992) view that comfort at home is achieved by a paradoxical combination of intimacy and control. As already mentioned, for women and children the notion of home can be perceived as a place of violence and horror precisely because it is considered a 'private' sphere and thus sometimes hidden from the eyes of public criticism and state laws. As we see the right to comfort at home is associated with other concepts of human and citizen rights related to freedom and prevention of violence, and sometimes the right to basic needs such as food, shelter, education, etc.

But there are other things that make people feel comfortable at home. For Mandy, who is a Canadian–Indian, comfort at home means the ability to talk in her mother tongue, which symbolises her ethnic identity. Mandy is a Canadian–Indian–Muslim who was born in Canada but lived at the time of interview in Jerusalem. Her mother tongue is spoken in her home in Canada where her parents live. In other places she speaks only English. Mandy's perception of home as speaking her mother tongue probably represents an immigrant perception of the notion of comfort at home and the deeper meanings of 'home' in its wider associations.

Many people mentioned home ownership as a major factor in the construction of a sense of comfort at home. Home ownership is associated with both the physical or material elements of comfort but also with the emotional–psychological side of being homeowners and feeling secure. Richard's definition of comfort at home illustrates these multi-layered meanings of comfort:

'It's mine . . . people are happy in my home, it's nice, it's a safe area and well looked after.' (Richard, 30s, married, British–Western, London, 1 September 1999)

'It's mine' is the important emphasis in this narrative. It's the notion of 'legal' belonging and ownership that provides people with the sense of security and comfort at home.

To conclude this section, the level of comfort at home reflects the gaps between the boundaries of one's own desires and aspirations, one's own cultural and gender identities and the realities of everyday life. The wider the gaps the less comfort one feels. These meanings of comfort at home appear to be universal. Most elements that are associated with the notion of comfort appear in most people's narrative, no matter whether they live in London or Jerusalem. Also, no difference was identified between people's different identities regarding comfort at home. Even people of 'the minority groups' in both cities described comfort at home with the same associations, especially with power relations. Nevertheless, the sense of freedom and the willing to bridge the gaps between the boundaries of one's own desires and their everyday realities are more explicit in women's narratives on comfort. In general, most narratives do refer to the three aspects, the physical, the social and the emotional, as important elements in making home comfortable.

Obviously the level and identification of comfort at home depends on ones' own financial ability, although none of the interviewees reflected a desire for a 'better' lifestyle as part of the identification of comfort. This is perhaps because most interviewees represent middle–upper-class status for which basic needs are more or less fulfilled.

Last but not least, it seems helpful to use the definition of 'comfort at home' as a basis for the understanding of 'a sense of home in the city'. These are sentiments and feelings of comfort, belonging and commitment that we feel in some parts of the city as much as we feel them in our home so that these parts are symbolically transformed into 'intimate spaces' where people can be 'themselves' as much as they are in their homes. These spaces could include areas of recreation within the city centre, cafés and small places with greenery where people can sit and pause. This idea that I wish to examine reflects a basic question on whether these elements of comfort, belonging and commitment at home can indeed become guidelines of urban planning and management.

Comfort in public spaces

Carr *et al.* (1992) identify comfort in public spaces as a basic human need but mostly in its technical, physical aspects: 'The need for food, drink, shelter from elements, or a place to rest when tired, all require some degree of comfort to be satisfied' (p. 92). They provide a checklist of the *physical* components of comfort in public spaces:

- relief from sun or access to sun
- shelter in general and especially from sun, rain or inclement weather
- comfortable and sufficient seating in public spaces
- the orientation of the seating
- their proximity to areas of access
- the availability of public toilets.

These are indeed the technical components of comfort mainly relevant to public parks in the city. Other researchers speak about social and psychological comfort as deriving out of a sense of security in a place, a feeling that one's own personal safety and possessions are not vulnerable. Safety is where one feels familiar with the environment and faces. A familiar environment brings relaxation as it diminishes anxiety (Fullilove 1996). Freedom of action, freedom from worry and concern also brings psychological comfort. How do these elements appear in people's own knowledge as related to public spaces? Let us analyse the meanings of comfort in the following categories: 'the building', 'the street', 'the neighbourhood', 'the city centre', 'the city', 'public transportation' and 'urban parks'.

The meaning of comfort in your building

'*The meaning of comfort in my building is similar to the meaning of comfort in my home; it is a feeling of space; hierarchy of spaces are important. I like order between spaces; I like different spaces for different activities (at home).*' (Janet, 40s, single, British – Jewish–English, London, 19 August 1999)

The building is the spatial unit, which comprises the home/flat. For some people such as Janet comfort in the building is like comfort at home, that is, they identify the building with the home/flat. For others the building is an 'in-between' space. As Janet mentions, there is a hierarchy of spaces; the building is not as 'private' as the home but it is not as 'public' as the street or the neighbourhood. Here too the level of comfort is affected by the boundaries between what people define as 'private' and 'public'. The fact that these boundaries are not so clear-cut causes discomfort to the residents of the building.

The building is the 'entrance' to one's own home. It reflects the changing symbolic meanings of space from the very 'public' to a 'more private' space in the sense that the building is less intimidating, more relaxed, more secured then 'public' spaces. Some find the corridor a 'strange space' in the sense that one passes through it and can smell cooking or hear voices or noises, private, intimate conversations or quarrels of other residents of the building, mostly strangers. It has a different kind of intimacy than at home, intimacy between strangers, which can cause a sense of discomfort.

Thus, if 'at home' the emotional aspects of comfort were dominant, 'in the building' the social aspects of comfort become the main issue because the interactions between 'us' and 'them' are more explicit and complicated in this category. As such it creates obvious ambivalent feelings among people as living in a building forces them to share common spaces and interests but at the same time to compromise on issues they feel are very intrusive to their own private space, such as smell and noise. Because the building is an ambiguous space where the boundaries between 'private' and 'public' are not always clear, a sense of discomfort is probably more common and it is usually associated with the forced social interactions with the neighbours that one has not chosen. A reflection of such ambiguous boundaries is expressed in Abraham's narrative:

> 'In the building the emphasis is on the cleanness of the building [as an element of comfort] because I could bump into strangers [neighbours] so at least the building's environment is clean and things are in their place.' (Abraham, 50s, Jewish, 29 February 2001)

Ronit's narrative expresses mixed feelings related to sense of comfort in her building, both physical comfort and social and emotional comfort. Ronit's narrative emphasises the physical aspects of comfort in the building, which are associated with its location. She lives in one of the low-income neighbourhoods in Jerusalem and perhaps this is one of the reasons why issues of maintenance and safety and security are highlighted in her narrative, although it repeats in other narratives as well:

> 'There is a lot of discomfort in the building, the stairs are not maintained properly, the building committee is not functioning so we don't have electricity all the time, the main door is broken, there is a harsh degradation in the building and its surroundings . . . but what is comfortable is that the neighbours are very warm people and if you are in trouble they help you out, there is somebody to listen to you, you are not alone . . . neighbours are better then family. (Ronit, 40s, divorced with three children, Mizrachi, Jerusalem, 8 June 2001)

What Ronit expresses in her narrative is the contrast between the physical discomfort and the social comfort. These contrasts are probably more significant to social interactions in lower-income neighbourhoods where strong social contacts minimise those ambivalent feelings between the physical proximity and the desired anonymity that people usually experience in their buildings. The building is for Ronit an extended family unit. She perceives neighbours as closer then family and this view makes the experience of living in the building much more positive and comfortable.

Ronit's attitude is an exception – for most people, both in London and Jerusalem, it is complicated to maintain good and distant relations perhaps because 'the building' is such an 'in-between' space, people are more cautious to keep their privacy. At the same time there are also those who wish to have neighbours similar to them so that they can become friends too (mostly people above the age of 50).

Unlike comfort 'at home', which was almost universally defined with neither differences between residents of London and Jerusalem nor differences between people of different identities, the meaning of comfort in the building involves or is affected by identity issues. Here notions of exclusion/inclusion are becoming significant. Thus for a Palestinian woman living in a building where all her neighbours are Jewish, comfort or discomfort means something else than for Jewish people living in the same building. For her comfort in the building is associated with her anonymity, the fact that the neighbours are not aware of her identity. Living in such a political situation her national identity becomes 'the problem' and she lives with the fear that the neighbours may discover that she is a Palestinian, 'an enemy'. She says:

> '*I would prefer that the building was similar to the city. I have chosen to live in Jerusalem because of its diversity, heterogeneity and its colour. The building is not like that. I am the only Palestinian in the building . . . I have my flat and that's it.*' (Aziza, 30s, single, Palestinian – Palestinian (citizen of Israel, Jerusalem, 7 August 2000)

Aziza's narrative reflects one of the dominant characters of Jerusalem – that of the segregation between Jewish and Palestinian neighbourhoods which exist in the city. There are hardly any buildings in which Jews and Palestinians live together. Aziza's situation is a rare example of a Palestinian woman who chooses to live in a Jewish-dominated building and neighbourhood, and she feels alienated because of that.

Obviously the boundaries of inclusion or exclusion depend on the type of identities at focus. For Moshe and Aaron, a homosexual couple, the meaning of 'comfort in the building' had the opposite connotation. They made sure that their neighbours knew that they are a homosexual couple and not just flat sharers because they felt that their sexual identity is accepted and perhaps also because they are Jewish, part of the majority. This is an example of emotional comfort, which is about the extent to which one's own identity can be made explicit and visible. Aziza, Moshe and Aaron can be labelled as 'the Other' but Aziza thinks its better to hide her identity and Moshe and Aaron feel comfortable enough to expose it. Obviously the choice of exposing one's own identity is a matter of one's assumption of whether this identity is accepted or rejected. In certain contexts homosexual otherness is much more accepted than national or Palestinian otherness.

To conclude, like 'the home' 'the building' is usually a homogeneous space where people of the same income level, culture and nationality live together but separately. It is here that the exclusion or fear of exclusion starts to take place and power relations become explicit, especially for those who are considered as 'the Other'. It seems that because of its ambiguous boundaries between 'private' and 'public' the building is one of the most complicated and contested spaces in which to feel comfort.

The meaning of comfort in your street

'*Comfort in my street is about its low density, which suits a family, its vegetation, its combination of privacy and commonality. I am a shared owner of the street with all the neighbours. I know all the 32 neighbours living in the street. We all manage and maintain the street. It's a middle-class suburban privacy.*' (John, 60s, married, British – white Anglo-Saxon, London, 11 August 1999)

John highlights in his narrative the connection between comfort and class. As he himself mentions he lives in a middle–upper-class suburban area in London where 'the street' is perhaps more private than in middle–lower-class areas in other parts of London. People in his area share the maintenance of their semi-private street. The street in such areas is part of one's own assets and its maintenance becomes a private rather then a public duty, partly because these streets do not serve 'the public' but only the residents of the area or their guests. The meanings of comfort in such a 'private' street in middle–upper-class areas is different than comfort in a 'public' street in middle–lower-class areas. In the latter case it is a street that one shares with many people with whom one has no direct contact unlike John's street, which is shared between well-known neighbours.

People both in London and Jerusalem emphasise the physical aspects of comfort in their own street:

- 'no traffic'
- small street ('so that I know the neighbours') – 'village-like' street
- location
- cleanliness
- greenery
- residential only (no shops or pubs)
- close to public facilities or having some facilities in the street
- a good proportion between the buildings (their height, width, distance between buildings)
- building style (small houses).

One of the respondents, an architect, describes a 'model of a comfortable street', a kind of checklist of the physical design he perceives as creating comfort:

'*Comfort in the street is about its* layout, *the* relationship between the building and the street, *its* width, *whether there is a feeling of* overcrowding or space. *If the* buildings are too much alike *it causes a sense of alienation, the* scale and the architectural variety *brings comfort.* Vegetation *brings comfort,* street furniture, street lights, *the* direction *of the street and its* relation to the sun *and* winds, *the* sounds in the street, *and the level of* air pollution. *All of these have their effects on creating a sense of comfort.'* (Emmanuel, 50s, married with three children, Israeli – Jewish, Jerusalem, 23 February 2000)

There are similar components that appear in both checklists – relationships between the buildings, width of the street, greenery or vegetation – as contributing to a sense of comfort. The architect adds several components that laypersons didn't mention, such as the direction of sunlight and wind or street furniture. Overall these two lists reflect a different and perhaps less dichotomous relationship between professional and local knowledge than that presented in Chapter 2. Perhaps these dichotomies are indeed less extreme when the planner and the community share the same cultural experiences and norms.

Comfort in the street is also associated with social aspects, although these aspects are much more elaborated on when people talked about their sense of belonging in the street (see Chapter 6). It seems that 'a sense of comfort in the street' is associated with physical aspects whereas 'a sense of belonging to the street' is associated with social aspects. However, those who did mention the social aspects of comfort in the street associated it with being friendly with people in their street:

> *'Comfort in the street means meeting familiar faces. When people walk relaxed, nothing threatening in their body language, cars move slowly, nobody is running at you.'* (Suna, 40s, single, Egyptian – Mediterranean, London, 29 July 1999)

Being friendly with the neighbours in the street is perceived as less intimidating and intrusive than being friendly with the neighbours in one's own building because it provides people with a sense of community and symbolises their belonging to this area but does not intrude into their privacy (elaborated on in Chapter 6).

'Friendly' – 'unfriendly' street

People made a distinction between their own street, where they have to go because this is the route to their home, and streets in general, where they go by choice. One woman made a distinction between residential streets and 'public streets'. Residential streets are not friendly for pedestrians she says, to the extent that one feels as alienated as if one gets into 'semi-private' space. What makes streets more friendly and perhaps more public is the functioning of public facilities, especially shops and cafés. These facilities make the stay of non-residents more acceptable and there is a feeling of inclusion among those who come to shop or sit in a café.

Street life is culturally constructed. Pitkin (1993) for example describes the function of the street in Sermontea, Italy after World War II as an extension of the house (see elaboration in Chapter 6). Indeed some people mentioned the street as a place of meeting. This is the case in the Palestinian village where Muhammad lives:

> *'The street is life. There is the street and there is the road. The road is for transport and the street is multi-functional. In the village the meaning of the street is a space of communication, people used to walk and meet other people. Today it is too dangerous because of the traffic.'* (Jerusalem, 23 April 2000)

Later in the conversation Muhammad explains that the social function of the street ceased to exist lately mainly because the street was appropriated by gangs of youth that threatened people. This description emphasises again the role of power and violence in appropriating the 'publicity' of the street. Here the powerful people are the youth who transform the street spaces in the Palestinian village into dangerous spaces and thus make people uncomfortable. Later in his story he mentions the dirt in the street and the garbage. What he describes is a situation that, unfortunately, is very common among non-Jewish villages and neighbourhoods in Jerusalem (see Chapter 4), where Palestinian villages and neighbourhoods suffer from lack of development and lack of municipal services such as rubbish collection in spite of the fact that they all pay municipal taxes. This situation can be defined as an abuse of citizen rights, the right to comfort in the street, which is based on a provision of basic infrastructure and services.

Fatma lives in another Arab–Palestinian village, which only lately had its infrastructure developed. In her narrative she makes an implicit connection between the notion of comfort in the street and the notion of 'the right to comfort':

> *'In my street it is comfortable because it is developed, there is a paved road, street lights so that I can enter my home comfortably, I have my private parking lot and the street has an "address". It is all developed physically.'* (Jerusalem, 18 April 2000)

For Fatma what makes her street physically comfortable is that it has the appropriate infrastructure, which enables her comfortable access to her home. These dimensions of physical comfort in the street were not mentioned at all by the Jewish people interviewed for this research, probably because these components are so obvious in their daily lives. Fatma mentions it because this new development of infrastructure changed her everyday life. Unlike John's 'private' street on the outskirts of London where maintenance is privately managed, in Jerusalem's outskirts it is part of the municipality's duty. Here forms of discrimination and oppression become more evident. The village where Fatma lives succeeded in being developed only recently and finally has the same level of infrastructure as Jewish villages had a long time ago (see Chapter 4).

Sense of comfort in the street has its commonalities in the two cities when referring to people of the majority. However, the fierce security situation in Jerusalem makes life in the street less comfortable. The Palestinians in particular, who are considered 'the enemy', are very much restricted in their movements in the street and the city by police control. The Jewish people's movements are also restricted because of the fear of bombs.

In general, we can conclude that comfort in the streets of Jerusalem is much less evident then comfort in the streets of London, mainly because of security circumstances that turn the streets of Jerusalem into a much more controlled space.

The meaning of comfort in your neighbourhood

Does the term 'neighbourhood' still make sense for a city's residents in the global era? Studies have dealt with the changing importance of 'planning by neighbourhoods' in urban planning in the last decades, arguing that 'planning by neighbourhoods' is back on the agenda of urban planning (Madanipour 2001). Other studies present a different argument of the loss of community and neighbourliness as one of the implications of the economic and social changes in global urban frameworks (Macnaghten *et al.* 1995). Lately, studies deal with the wider global trends towards gated developments and communities that will grow in the future (Webster 2001). How do the residents of London and Jerusalem feel about their neighbourhood? Richard's narrative regarding comfort in his neighbourhood presents some of the current perceptions of the notion of neighbourhood:

> 'Neighbourhood for me means rows and rows of houses. Comfort in the neighbourhood doesn't have any meaning for me. Comfort at home does have, but comfort in the neighbourhood doesn't have. In my childhood it did have a meaning ... my parent's generation found it easier to communicate but now, our house could be anywhere ... perhaps it's because we don't have children.'
> (Richard, 30s, married, British – Westerner–English, London, 1 September 1999)

In this research the loss of sense of neighbourhood is also expressed in the fact that people could not easily define the boundaries of the neighbourhood, and thus the meaning of 'comfort in the neighbourhood' was unclear to many of them.

For most people in both cities the identification of the neighbourhood's boundaries imply the areas they mostly use, that is, shopping areas. For those who have children the boundaries of the neighbourhood are where schools, kindergartens and other such facilities are located. Those who don't have children, such as Richard, are less aware of specific neighbourhood boundaries.

For people living in London big roads define neighbourhood's boundaries. For some in London and Jerusalem the neighbourhood is the street they live in and therefore comfort in the neighbourhood means comfort in their street. For others the neighbourhood is the high street (Baqa'a in Jerusalem) or the whole town (Kew or the East End in London). Sometimes the boundaries of comfort in the neighbourhood are drawn by the locations of shops and shopkeepers whom people know in person.

It appears that sometimes a sense of neighbourhood is connected to one's own identity and one's own stage in the life cycle. This is very clear for mothers, for whom a sense of neighbourhood became stronger when they had small children and they used the neighbourhood facilities more intensively (see Chapter 8 for elaboration).

Another gender dimension, which becomes part of defining one's own comfort and constructing the spatial boundaries of the neighbourhood, is women's clothing. In Jerusalem the cultural differences between neighbourhoods are so contrasting that women's clothes become a sign of the boundary marker between the 'forbidden' and 'the permitted' (see also Chapters 3 and 8). Mandy, who lived at the time of the research in Jerusalem, made this connection between comfort in the neighbourhood and her clothing:

> 'In the summer I don't like to wear shorts and go into religious neighbourhoods located next to my neighbourhood.' (28, single, Canadian – Indian–Muslim, Jerusalem, 16 June 2000)

Mandy would bypass those neighbourhoods, 'forbidden' for her in many ways just because her clothing is considered 'inappropriate'. This notion of women's clothing and the freedom of movement in Jerusalem repeats itself for women of different identities (both Jewish and Palestinians), as will be shown in Chapter 8. There is another aspect of clothing that marks the boundaries of the neighbourhood, not only for women but also for men. It is what Aaron terms 'the boundaries of the tracksuit', that is, the extent to which he feels comfortable wearing his tracksuit in the street. Here the boundaries of the neighbourhood determine another dimension in the private–public dichotomy – Aaron feels 'at home' in his neighbourhood so he wears casual clothes that he would wear at home.

Neighbourhoods usually reflect economic, cultural, religious and ethnic homogeneity. People tend to live in vicinity to people similar to them. *'The East End feels like you're not in England but the system is English'* says Gaziul (30s, married, British – Asian–Bengali, London, 7 August 2001). Gaziul has his own business in the Spitalfields area in East London and lives in North London in an area where most of his neighbours are also Bangladeshi. Most of the Bangladeshi people who live in the Spitalfields area said that for them Spitalfields is like 'back home' in the sense that the social contacts they have and the goods and food they can find there is identical to what they had at home (see Chapter 4). It is where they can speak their mother tongue and where the street names and shop signs are marked in their mother tongue. For most of them 'the neighbourhood' is their 'home'. It is the place with the elements that are associated with home in London – visual signs, smells, music and clothes. Their neighbourhood becomes their social and cultural network, which is so crucial for their everyday life in London. This is also true for some English whites: *'The neighbourhood is where you've got your roots'* (Eleonore, single, 50, British – white, London, 1 September 1999). Comfort here is linked to notions of belonging, of being rooted.

In Jerusalem, most neighbourhoods are homogeneous on a national and religious basis. As mentioned before hardly any mixed neighbourhoods of Jews and Palestinians exist in the city and the same is true for mixed neighbourhoods of secular and ultra-orthodox Jews. Thus Neta mentions that she feels comfortable in her neighbourhood because only secular Jewish people live in it: *'their norms are similar to mine. I also feel comfortable that there aren't many rich people – no snobs living in my neighbourhood'* (50s, married with three children, Israeli – Ashkenazi, Jerusalem, 11 June 2000).

It seems that in both cities people feel comfortable living in neighbourhoods with people with similar identities to themselves, whether they are Bangladeshi in London or secular Jews in Jerusalem. Sometimes the 'others' can feel comfortable as well, depending on the level of openness that the majority possess towards those 'others'. Aziza's experience as a Palestinian woman who lives in a Jewish neighbourhood is quite positive, although when she discussed comfort in the building her experience was not so positive. Perhaps the differences in her sense of comfort in the building and in the neighbourhood rely on the fact that the neighbourhood is located very close to the borderline between West and East Jerusalem:

> *'I feel very comfortable in this neighbourhood. It is the most beautiful place in Jerusalem . . . I am a woman with constraints: I am a Palestinian, single, and this is the ideal neighbourhood for me – a microcosm that reminds me of London . . . a little bit of everything. For example I meet some Palestinians [who come here to sit in the cafés]. I also see Chinese people and religious Jewish people. Here is my home and I flourish in places like that. It is on the borderline between East and West and it is ideal for me. Previously when I lived in Rehavia [a Jewish neighbourhood] I felt suffocated.'* (Aziza, 30s, single, Palestinian – Palestinian (citizen of Israel), Jerusalem, 7 August 2000)

However, as mentioned before, Aziza's experience is unique. Palestinians (even Israeli citizens) do not tend to live in Jewish neighbourhoods either because they fear being excluded or intimidated or because they themselves prefer to live among other Palestinians, even at the cost of living at a lower standard.

Comfort in the neighbourhood is associated with knowing people, encountering familiar faces. Eitan describes the intricacies between scale and intimacy:

> *'The larger the scale the more it is important for me to feel intimacy. At home I look for an ultimate privacy, that is, not knowing neighbours, but in the street [and neighbourhood] it is important for me to meet familiar faces.'* (Eitan, 36, married, Israeli – Mizrachi, Jerusalem, 3 February 2000)

The neighbourhood and the street represent a clear-cut and less ambiguous dichotomy between private–public boundaries than the building and therefore for Eitan the meaning of comfort in the street and neighbourhood, that is, knowing people, is precisely what makes him feel discomfort in his building because it threatens his privacy.

Social support system is another way of expressing the social meanings of comfort in the neighbourhood says Efrat, who is in her 40s, married with two children, an Israeli (Jewish): *'There is a lot of mutual help in the neighbour-hood, we share the children's transport and other mutual help.'* Perhaps mutual help in the neighbourhood is also a matter of stage in the life cycle and it is relevant mainly for women with small children (elaborated on in Chapter 8).

Physical aspects of comfort in the neighbourhood both in London and Jerusalem are associated with:

- facilities in the neighbourhood such as clubs, cafés, green areas, schools, community centres (especially in low-income neighbourhoods)
- 'physical harmony' – that there are reasonable relationships between the size of the buildings
- parking spaces
- not too many religious places (for example, synagogues in Jerusalem).

To conclude, the notion of 'neighbourhood' has become irrelevant for some people in both Jerusalem and London mainly because the boundaries of the neighbourhood are not so clear to them in terms of the services they use. Here identity issues matter in defining comfort in the neighbourhood. It appears that for people of low income level or people of specific ethnic or cultural settings such as the Bangladeshi in East London the notion of neighbourhood is more meaningful in the sense of social ties, mutual support system, social and cultural facilities, etc., than for English white people, especially those with no children. Also, a sense of neighbourhood is more meaningful for mothers who are in charge of child rearing and thus 'use' facilities such as schools, which for them define the boundaries of the neighbourhood.

Comfort in the city and the city centre

'I love the city centre. For me the city centre is the heart of the city where life, business, shopping and everything else happens.' (Elizabeth, 50s, single, British – Israeli–Jewish, 16 March 2000)

'I try to avoid the city centre, it's too noisy and too busy. I wouldn't go just to go. I go only for a purpose (for example to Oxford St.). I don't like shopping any-way.' (Lynne, 40s, single, Scottish – white–Western, London, 22 August 1999)

Two expressions towards the city centre. The two relate to both cities. In both of them some people liked the city and its centre and others did not for various reasons.

People use the same associations and symbols to describe their fascination with the city centre in both cases: 'exciting', 'diverse', 'full of attraction', 'diver-sity of colours', 'food', 'dresses', 'religion', 'choices', 'art galleries', 'cinemas', 'a lot of cafés', 'surprises', 'cosmopolitan', 'multicultural'. Even what makes people feel discomfort in the city centre is similar in the two cities; 'pollution', 'noise', 'dirt', 'overcrowding', 'intimidating', 'appropriated by gangs of youth'. It seems that both city centres suffer from the same urban maladies, which probably characterise other cities as well. Problems of air pollution, which London is mostly known for (see Chapter 4), problems of cleanliness, over-crowding, etc., which the two cities are known for. What is perhaps more surprising is that residents of Jerusalem use the same associations to describe what makes them feel comfortable in the city centre – terms such as 'cos-mopolitanism' are used in both cities.

What is more common to both city centres is the role of the café in city life:

'A city without cafés is not a comfortable city. Without cafés a city centre is not a real centre and a city is not a city! Cafés symbolise a pause in a constant, vibrant, non-stop movement, which is what the city is for me . . . in a café one can sit and gaze at people. For me this is the fun.' (Linda, 30s, single, Israeli – Western, living in London, London, 30 July 1999)

'Cafés are what make a city become a real city. The more they exist the more city-like a place becomes. A new café is like a new flower.' (Aziza, 30s, Palestinian – Palestinian, Jerusalem, 7 August 2000)

The café means a pause, a break between activities. It represents spaces in between the public and the semi-private where one can stop, a point of relaxa-tion and reflection in the midst of the rush of the city. The café legitimises the gaze, the eye contact intrusion, which reflects boundary crossing as well. The café sitters become the legitimised audience in the theatre of everyday life.

The café is one of the expressions of modernisation and urbanisation. It is one of the later outcomes of the industrial Revolution and the expansion of cities. It began to appear in cities in the mid-nineteenth century (Wilson 1991). The café is a place where culture, philosophy and art become explicit. From the Impressionist painters in Paris or Simone de Beauvoir, whose 'personal drama of her intellectual and sexual emancipation seemed inseparable from its setting in the hotels and cafés' (Wilson 1991: 63) to Walter Benjamin in Berlin – all used the café as their place of creativity and inspirations. From the mid-nineteenth century onwards the café became a central cultural phenomenon, a place of pleasure, and a place where personal boundary crossing could take place because norms and values were becoming somehow vague. Thus the café is where one can get drunk, tell dirty jokes, stay until late at night, meet new people, start or end relationships. The café was not only a place for intellec-tuals. Wilson (1991) notes that the café was a place to go out with the whole family for the bourgeoisie in Paris in the mid-nineteenth century since their apartments were overcrowded. Some thought that because of the café 'the home is dying' (Wilson 1991: 52).

The café as an icon of urbanism and city life is still a very dominant com-ponent, which makes people feel comfort and belonging in both London and Jerusalem. The nature of café entertainment is different in the two cities

Figure 5.1   A café in Jerusalem

primarily because of the different types of weather, which dictate whether the café sitting is an indoor or an outdoor activity. In Jerusalem, café sitting is an outdoor activity most of the year (see Figure 5.1), whereas in London it is mostly an indoor activity except in summertime (see Figure 5.2). Still, in spite of the differences in their spatial activities (indoor/outdoor activities) they serve the same purpose as an icon of urban activities in these two cities and probably in cities in general.

Spatial order in the city has been mentioned as another dimension that increases a sense of comfort in the city and the city centre. Spatial order means a clear division of the city's activities between spaces where people walk, where cars drive, and where people do their shopping or enjoy their recreational activities. This urban order argues Abraham – himself an architect – is what provides a sense of comfort in the city and city centre. It is no coincidence that urban order as increasing a sense of comfort has been expressed by an architect. Urban order has been the essence and the primary rationale of modern urban planning since the beginning of the nineteenth century. Over the years it has been used as a motive to justify many of the maladies of modern cities, including urban rehabilitation and neighbourhood destructions.

Let us return to the notion of 'the right to comfort' as it is expressed in daily practices in the two cities in order to incorporate people's narratives within the language of citizen's rights. The right to comfort is highly connected to the political context in which the city and the state function. The attitudes to

Figure 5.2   Film Café at the British Film Institute, South Bank, London, in summer

minorities living in the two cities is an expression of nationalistic policies of exclusion and discrimination, as is the case in Jerusalem, or capitalist policies that find their expressions, especially in the East End of London where the Bangladeshi people live (largely elaborated on in Chapter 4). The Bangladeshi living in London have split feelings towards the city. The older generation take the city centre as an overcrowded, noisy place and therefore some of them do not go into the city centre at all. The young generation find the city centre and the city very fast but they like its diversity and openness:

> *'I can live only in London because London as compared to other cities in Europe and America is the most open and the most warm to other cultures. Not in Britain – just in London!'* (Suliman, 30s, married, Indian – Bengali, London, 31 August 2001)

Suliman carries on and explains that he feels comfortable in London. He is a 'city person' and he doesn't feel like a foreigner at all. London is a 'no one area', as he terms it. What he refers to is the sense of anonymity and acceptance that he feels in London. His narrative does not mention at all the developers' intentions to build new office buildings in the area. When I asked him he said that he was not familiar with these projects – perhaps because he does not live in the area. For Suliman the right to comfort in London seems to be fulfilled. Amal's narrative expresses a more explicit sense of discomfort and abuse of her right to comfort that most Palestinians feel in Jerusalem, a sense of discomfort that is a direct result of the politics of discrimination and exclusion towards Palestinians living in Jerusalem:

> 'East Jerusalem city centre is dirty! Full of spitting, stinks from urine, full of soldiers and policemen – "intimidating people", a lot of unemployed youngsters . . . West city centre is really the opposite . . . free behaviour, nobody knows you. I can sit in a coffee shop, walk around with anybody, act freely . . . unless you have a Middle Eastern look, then they [the soldiers] chase you . . . they don't do it to me because I don't look like an Arab.' (Amal, 40s, married with two children, Palestinian – Arab–Palestinian, Jerusalem, 7 August 2000)

Amal's narrative is a reflection of the nationalistic politics of planning and development in Jerusalem. These policies create significant differences between the level of development and the provision of municipal services in West and East city centres and these differences are expressed in people's narratives. There are also significant differences in the degree of control and surveillance in the two sides of the city, reflecting the different recognition of 'the right to comfort' in East and West Jerusalem. In East Jerusalem 'the right to comfort' is constantly abused whereas the centre in the West side of the city seems from Amal's narrative more like what a city centre should provide.

It must be noted that the interviews with the Palestinians in Jerusalem took place just before the second Intifada started (in October 2000). Since then the situation has deteriorated tremendously in terms of the level of control and the police and army's visibility in the streets of both West and East centres of Jerusalem. It is most probable that Amal's narrative now would be different, as she would not find it so comfortable to go to West Jerusalem, as much as Jewish people don't go these days to East Jerusalem because of the increasing security problems in the area. These expressions of the abuse of the right to comfort in the city repeat themselves in all the interviews carried out with the Palestinians. Muhammad talks about his experiences:

> 'I don't feel comfortable in East Jerusalem because I see the youth gangs and the provocations and the crowds of people and the tension in the street . . . policemen . . . this is not a city . . . these are people on drugs . . . and soldiers . . . I get into East Jerusalem and see the Israeli flag everywhere as an act of power . . . in the West I don't see soldiers, not in the same number as in the East. So it's the

political, religious and social factors that make the difference between East and West city centres and the dirt and crowds as well .' (Muhammad, 30s, married with one child, Palestinian (citizen of Israel, Jerusalem, 23 April 2000)

What the Israeli government and municipality perceive as a 'united city' is in fact a divided city. The two city centres are perceived as 'forbidden' and as 'dangerous places' for the other population. For most Jewish people the East side of Jerusalem has become 'non-existent' in terms of shopping or outings since the violence started in October 2000. Most of them do not use it, mainly because of security tensions. For the Palestinians living in the East side, the West city centre is far from being democratic, equal and exciting. None of those centres thus provide 'the right to comfort', either for the Jews or for the Palestinians. The city centres are very different from the democratic, equal, sharing and character of London. It is rather the opposite – the city centres on both sides are very much controlled and inspected, mainly towards the Palestinians but towards Jews as well.

To conclude this section, the analysis of people's narratives on comfort in the city centre reveals the wide range of perceptions that people have regarding their city centre. Some of these perceptions are similar to people living in both Jerusalem and in London, especially those related to the essence of city life such as the importance and significance of the café, the city's attraction, its diversity, etc. The main differences between the two cities lie in the different power and control relationships that exist in Jerusalem, which is more significant than those in London, although in both cities the politics of planning and development reflect mechanisms of power – nationalistic in Jerusalem and capitalistic in London. However, from people's narratives it seems that Jerusalem is a much more controlled city, a city of fear and terror, while London is becoming more of the 'city for all' in spite of the powers of capitalism that harm the 'have-nots'.

A sense of comfort on public transport

'What I like about public transport in London is that it's modern and efficient. I don't like run-down, old transport. I don't feel confident. I don't like when trains are full of people's food leftovers. I don't like empty trains. I don't like low ceilings. I'd like to have a lot of room for my legs and bags (especially in rush hours). I don't like hot tubes.' (Stuart, 30s, single, British – Ghanaian–Afro-Caribbean, London, 6 August 1999)

Public transport is one of the most basic components for maintaining city life routine. The use of public transport has become widespread in the last decade even among car-owners, since access to city centres with private cars has become impossible because there are just not enough parking spaces in the city centre and because of the daily traffic jams in peak hours. Stuart's narrative reflects the main advantages and disadvantages of using public transport in London. It is a multi-layered system: underground, buses, trains, taxis and minicabs (which are a kind of public–private system). Remember the figure of over 3 millions journeys a day that are made on the tube in London (see Chapter 4). This is a significant number of journeys, which necessitate a very efficient system. What Stuart describes are some of the most common complaints of tube travellers regarding the experience of using the train and tube system, such as James's observation that the system '*is inefficient, alienating, inaccessible . . . total impotence.*' (James, 60, single, British – Jewish, London, 22 August 1999). Suna (40s, single, Egyptian – Mediterranean) does not like the intimidatory aspect of the journey, especially by '*gangs of kids, people that are drunk, especially after football matches*' (London, 29 July 1999).

As a woman she feels intimidated by carriages full of men or empty train carriages at night. The latter cause a sense of discomfort for men too, such as Stuart, as he said in his narrative above. The practice of power is expressed very clearly in travelling in public transport in London, especially during 'problem' periods such as at night, early in the morning or on special occasions such as after football matches. It is less so in buses as the presence of the bus driver makes the whole journey more comfortable, even late at night.

In Jerusalem, public transport consists of buses and taxi services. (At the time of writing there is no train system within the city or between Jerusalem and other cities in Israel. There are plans to construct railway services to replace the old system but these have not materialised). People living in Jerusalem complain about the lack of trustworthy timetables, lots of delays, a network that is not efficient enough, etc. Those who have cars prefer to use them. Nevertheless the fear of terrorism and suicide bombers on buses prevent many people from using buses in Jerusalem and those who use them, both Jewish and Palestinian, feel very tense. A period of 'urban terrorism' is reflected in the increasing feelings of fear and tension among Jews and Palestinians alike as they become targets for bombing, as they all use buses in their daily lives in West Jerusalem. For Palestinians this period is perhaps more intimidating as they are constantly associated with 'the enemy'. Perhaps the most explicit expression of these tensions is manifested in Fatma's daily practices of using the bus:

> '*When the second* Intifada *[Palestinian uprising] started I used to turn off my cellular phone before getting on the bus because I knew that if something happened and the people heard it in the radio and they heard me speaking Arabic on the phone . . . I am not sure what would have happened. At the same time, it is quite comfortable to use public transport. I get where I want and I use it quite a lot as I don't have my own car.*' (Fatma, 40s, single, Palestinian – citizen of Israel, Jerusalem, 18 April 2000)

Fatma would not use her mother tongue Arabic when there is a situation of increased tension in the city because she fears that the Jewish people might associate her with 'part of the enemy'. Previously we mentioned how Mandy related to speaking her mother tongue as an expression of home and here Fatma prefers not to speak her mother tongue in public because she fears the reaction of the Jewish people.

'For me as a Palestinian woman there is a problem of connection between the West and the East side of the city. If I want to use public transport to go to East Jerusalem I can't. There are no regular buses that connect the two sides of the city – the transport system is an indication of apartheid in the city.' (30s, single, Palestinian – citizen of Israel, Jerusalem, 7 August 2000)

The public transport system is one of the explicit expressions of the de facto division of Jerusalem. There are two separate bus systems that function each in another part of the city. There are routes with the same bus numbers going in different ways so that for example bus number 18 in West Jerusalem goes to other destinations from bus number 18 in East Jerusalem. How can one tell the difference between the two lines? By the physical condition of each bus, by the Arab or Hebrew signs on them, by their different colours and by the people's appearance in the buses. This separate system is clear spatial evidence of how the city functions as a divided entity that is in contrast to what the city and government politicians try to present – a united city.

Comfort in urban parks

Most people living in London expressed their positive perceptions of London's parks. 'There is a variety of trees and flowers', 'good views and vistas', 'big sizes of parks', 'good landscape planning', 'wonderful', 'exciting', but also 'too manicured', 'too crowded'. Richard elaborates on his positive attitude towards the parks:

'I always feel comfortable in open spaces . . . away from buildings and being able to see some greenery, especially in Britain. It's good to have some green areas.' (Richard, 30s, married, British – Western–English, London, 1 September 1999)

Parks symbolise 'nature' in the city. They are there for us, the city citizens, to be able to have a break, a pause in the rush of city life. To have a space of greenery, trees, flowers, as if to remind us what 'nature' means as opposed to 'city'. Parks are one of the key land uses in cities in the modern world and their importance has made them one of the key land uses in urban planning.

Urban parks have a peculiar character that distinguishes their use and perception from other 'public spaces' (such as city centres, streets, etc.). They serve as 'public' spaces but they are perceived by their users as 'semi-private' spaces in the sense that there are certain rules of behaviour which are required of park users and not for example of street or city centre users: 'no cycling', 'no barbecues', 'no dogs', 'no ball games on the grass', 'keep off the grass', 'no fires' (see Figures 5.3 and 5.4) are only some of the restrictions that exist in urban parks in the two cities.

People with whom I talked about comfort in urban parks said that comfort is associated with populated parks, preferably with people of similar identities,

Figure 5.3 Park regulations in Jerusalem:
Keep quiet between 14:00–16:00. Do not walk dogs without a lead. It is totally forbidden to light fires in the park area. Children under 4 must be accompanied by adults.

but not too many people, because comfort in the park is also associated with privacy:

> *'Sparseness of people, no overcrowding, in the street I don't mind overcrowding, in the park I mind! . . . I want to feel like I am in a "bubble", that I talk to someone and no one hears me or that I don't hear the neighbour . . . that people speak in my language, this is comfort for me, that I can talk to people and they understand me. Home and park are somewhat similar in this sense.'* (Abraham, 50s, married with three children, Jewish, 29 February 2001)

Stuart, who lives in London, expresses his own perceptions of 'comfort in urban parks':

Figure 5.4   Park regulations in London:
These regulations have been drawn up to protect visitors to the Royal Parks as much as the Royal Parks themselves. We encourage the users to enjoy them to the full. Abuse of the Park and its facilities will, however, result in penalties.

'When there aren't too many people, when there are a variety of things – flowers, hills, trees, viewpoints where you can see the city and see the relations between the city and the natural environment. Variety of use. I hate small parks, they are depressing, they are so small and very vulnerable, less well kept, get homeless people and they are not safe, not cleaned.' (Stuart, 30s, single, British – Ghanaian–Afro-Caribbean, London, 6 August 1999)

Several issues emerge from Stuart's narrative. First, the desire that the 'public' will function as 'private' too, perhaps in terms of 'the appropriate behaviour' as their rules and restrictions indicate. Stuart also mentions his expectations of cleanliness and good maintenance of urban parks. Both Stuart and Abraham mention the desire that other park users will be similar to them in the language they use, in their appearance and socio-economic status. Stuart does not like to meet homeless people in the park. For him they become 'the Other', those that people feel unsafe among. In London the 'others' are also people of different socio-economic and cultural background. In Jerusalem, 'the Other' can be people of different religious orientation (ultra-orthodox) or people from a different nationality (the Palestinians).

Since the number of parks in East Jerusalem is very small, Palestinians use urban parks on the West side of the city. However they use only those parks that are located on the borderline between East and West Jerusalem. It is interesting to note that the two populations (Jewish and Palestinian) do not mix and each has its 'appropriated' space in the park. These territories are clearly known to the park users, although not marked physically. Several researchers have dealt with the question of whether urban parks that are located between racially and economically different neighbourhoods can become 'green walls' or 'green magnets', suggesting different ways of measuring the 'wall effect' and the 'magnet effect' (Gobster 1998; Solecki and Welch 1995). It seems that from people's everyday experiences and narratives most parks in both London and Jerusalem are perceived more as green walls than green magnets (see Figs. 5.5 and 5.6).

The park is perceived by people of both Jerusalem and London as a sanctuary, as aesthetic, as a place of memory. For those who use the park for sport it has become a space of 'communities of users': walkers, sports people and dog

Figure 5.5 'Appropriation' of park areas in Jerusalem

Figure 5.6 'Appropriation' of park areas in London

walkers each develop their own community culture of daily meetings in the park on the base of a common, shared interest.

For women the use of the park is not so peaceful:

> *The park is a scary place, I avoid it when walking to work. In the morning it is empty and men . . . intimidate women.'* (Rebecca, 36, married, Israeli – Ashkenazi, Jerusalem, 3 February 2000)

The very same character that distinguishes parks from other public spaces in the city, their 'public–private' nature, is what makes them intimidating for women. Much has been written about public parks and fear (Madge 1997; Gobster 1998; Solecki and Welch 1995). Most works connect the use of urban parks and 'the geography of fear', especially for women, elderly people and Asian and African-Caribbean people (Madge 1997). Fear is one of the leading constraints in the use of urban parks for both women and men, although it is a more dominant constraint for women. In their narratives on 'comfort in the park' mostly women used fear as causing discomfort when using parks, especially at nights. Most women, both in London and Jerusalem, avoid using urban parks most of the time during daylight.

To conclude, urban parks represent another 'private–public' type of space in the city. Their importance as a place of sanctuary and relaxation makes them be a 'semi-private' kind of environment, with specific restrictions that distinguish them from other public spaces such as streets.

Terror in the city and other public spaces

It is almost impossible to discuss notions of comfort in the city and other public spaces without coming across the issue of terrorism and its effects on people's daily lives. Currently, daily life in Jerusalem is much more affected by terrorism, but terrorism had its tremendous effects on Londoners during the 1990s as well. In fact it seems that in the last few years terrorism is becoming more and more associated with city life. This is because national and political tensions and conflicts have their very explicit expressions and effects on cities' function or disfunction these days. 'Urban terrorism' has its very clear geographies. The twin towers attack on 11 September 2001 emphasised the symbolism engaged in the selection of sites for attacks. In New York it was the representation of the Western world of capitalism – the World Trade Centre. During the 1990s the IRA attacked banks within the City of London, which signify both its imperialistic role and its role as a world city in new Europe (Jacobs 1996). In Jerusalem, the selection of sites for attacks is targeted to destruct everyday and routine life of the city citizens rather then blowing up symbolic sites. This is because urban terrorism in Jerusalem has as its goal the transformation of 'public spaces' into 'dangerous spaces' and thus making people feel very uncomfortable and unsafe wherever they move in the city. Because terrorism in Jerusalem happens everywhere it affects the very basic essence of city life: the freedom of choice of where to eat, drink, shop, stop at a café. It also affects the anonymity of wandering around where one feels like in the city. In the attack on the Twin Towers in New York some 2,801 people were killed, five times more than in terrorist attacks in Israel in the last two years (end of September 2000–end of July 2002). But it seems that terrorism has had an enormous effect on Jerusalem's city life, especially in the last two years. At the time of writing Jerusalem is one of the target cities of suicide bombers in Israel. In the period between 29 September 2000 and 31 July 2002 there were 224 terrorist attacks in Israel, out of which 142 (63 per cent) occurred in the occupied territories, 54 (24 per cent) occurred in cities or other regions in Israel and 28 (13 per cent) occurred in Jerusalem and its outskirts (MFA 2002). 586 people were killed (both soldiers and civilians) and 4,204 were wounded. There is no other city in Israel that has suffered from so many attacks, although terrorist attacks take place in many other cities (such as Tel Aviv, Herzelia, Kfar Saba, Hadera, Haifa, Netanya). Even so, it seems that in Jerusalem the intensity and centrality of the attacks create a sense of constant discomfort and fear more than in any other cities in Israel.

Terror in the city has thus been an issue that most people living in Jerusalem raised in their discussions of comfort in the city. Mary's narrative is one such example:

> 'What is not comfortable in the city centre is the constant sense of tension that I feel because of the bombs, for example I don't go to the market on Fridays . . . I just can't because there were bombs in many places in the city and there are a lot

of "memorial sites" that don't let you forget. Also the soldiers there, on the one hand I feel secure, on the other hand it is disturbing.' (Mary, 40s, single, Israeli – Ashkenazi–Jewish, Jerusalem, 4 August 1999)

Terrorism in Jerusalem happens in places that represent the city's crowdedness – in buses, in restaurants, train stations, shopping malls, parks, supermarkets, markets, bus stops and cafés. Its unpredictable nature is what terrorises and prevents people from capturing the essence of city life. Out of fear people begin to draw their own 'forbidden' and 'permitted' mental maps of the city and this fear of terrorism is what brings discomfort and actually the disfunctioning of the city centre and the city as a whole.

London suffered from IRA terrorism as well, especially during the 1980s and the 1990s. These attacks caused many casualties and during that period there was a fear that terrorism would also affect urban life. When terrorism was at its highest intensity in London it increased the level of control and surveillance in public spaces in the city. If one reads the London daily newspapers published around the end of 1996 one can observe texts similar to those currently appearing in Israeli newspapers. For example, on 6 December 1996 in an article entitled 'London's ring of steel to expand' the enlargement of the 'ring of steel' security zone to the West of the city was reported. The article also mentioned the extension of traffic restrictions and police checkpoints that were set up after the IRA bomb attack on Bishopsgate in February 1993 and showed a map of the existing security zone and its extension, which actually covers most of 'the City' area.

Davis (1990) in his analysis of the fortress city emphasises the extent to which militarisation of urban life is a middle-class response to racialist or criminalist differentiations produced in cities by uneven economic and political restructurings. In Los Angeles, where Davis's work is based, these forms of militarisation are expressed in the practice of social cleansing and segregation. In Jacobs' analysis of London (1996) the 'ring of steel' and its camera watch surveillance system has been 'a strategic response to a residual colonial predicament. It was especially directed to keep the IRA outside the City' (p. 67). But as Jacobs argues it was not only to 'purify' the space but rather a strategy to obtain and guard its position as a 'global city'.

In Jerusalem militarisation and surveillance have existed since Israel's establishment in 1948 but have increased tremendously since the start of the last Intifada. Surveillance and control are much more explicit and visible in the city and they become mechanisms that work to prevent the Palestinian terrorists from committing their suicide bombings, but these procedures have their effect on every Palestinian who works or uses the city (especially in West Jerusalem) because they cause the long queues that the Palestinians have to face at army check points.

Summary and conclusions

Analysing the meanings of comfort in people's daily lives helps us to understand the deeper meanings of other processes that shape our everyday lives, such as the role of power relations and the boundaries between private and public which emerged from people's reflections on comfort. Analysis of this notion exposes the rich associations and implications that cut across the different hierarchies of space. Here are some insights from the analysis:

- *The dominance of power relations* both at home and in public spaces is a major trigger affecting people's daily practices of comfort. Comfort at home is mostly associated with the power of the patriarchy, especially for women. Thus the level of comfort at home reflects the gaps between one's own desires and aspirations and the practical ability to realise them. In public spaces the association of comfort with power relations relates to the role of the authority (state and municipality) in developing the basic infrastructure and services that bring physical comfort. The dominance of power relations in dictating comfort is also linked to public spaces, appropriation by youths in Palestinian streets or mobs on tubes in London.

- *Comfort and the boundaries between 'private' and 'public' and the spaces in between them.* The discussion on comfort exposes the large variety of symbolic spaces that exist between the two well-known and largely discussed ones – the 'private' and the 'public'. The discussion on comfort at home shows that even within the home, which is usually considered the most private space, there are multiple layers of 'private' spaces. At the home level this classification is gendered and perhaps also age oriented. It also depends on the level of freedom and space that one feels in one's home. It repeats itself in the building, which represents in many ways a confusing space in terms of the boundaries between the private and the public, and also in urban parks. Such confusions and dialectics are perhaps the main causes of a sense of discomfort in the building and sometimes in parks. In addition, several functions in the city such as cafés and parks represent 'semi-private' spaces within the public one in the sense of their associations with comfort and privacy.

- *The dominance of the symbolic boundaries between 'me'/'you' and 'us'/ 'them' in determining a sense of comfort or discomfort.* The determination of comfort is in many cases a matter of boundaries allocation, not only between the 'private' and the 'public' but also between 'us' and 'them'. When these boundaries are not clearly identified, such as in buildings, it brings discomfort and a desire 'to keep one's distance' from neighbours, whereas when these boundaries are more clear-cut, for example in the street, comfort becomes a desire for social encounters with neighbours. These boundaries between 'us' and 'them' are what determine norms of exclusion and inclusion both formal, that is, based on official legislation, or informal – expressed in human communication and behaviour.

- *Comfort as a gendered issue* is one conclusion that emerges from the above observations. It seems that for women the notion of comfort at all levels of space is more concerned with power relations (elaborated on in Chapter 8). For women comfort and discomfort results from a sense of security or fear, for example in the use of urban parks in the city. Because of its articulation with power relations, private/public sensitivities and boundary issues, comfort contains specific meanings for men and women, which differ sometimes on the basis of different experiences and cultural norms.

- *Comfort as a class issue.* Class and lifestyle have a significant effect on one's own definitions of physical comfort, not only at home but also in public spaces, especially in the street. In the narratives presented in the chapter this point was strongly emphasised when different kinds of lifestyle reflected different definitions of 'comfort in the street', especially in London where in suburban areas streets are 'semi-private' and are maintained and managed by the residents themselves – they are thus perceived as more comfortable.

- *Comfort and social, cultural and ethnic homogeneity.* People tend to feel more comfortable with people who are similar to them. This point was raised when the notion of comfort in the neighbourhood was at stake. In residential areas comfort is about social, cultural and ethnic homogeneity, which is in contrast to comfort in the city centre. This type of homogeneity can lead to different levels of 'ethnic towns' such as Chinatown or Banglatown. It seems that people prefer to live in homogenous neighbourhoods especially when ethnic or religious customs are more dominant in the community's life. Thus a neighbourhood such as Banglatown functions to serve the special food, music and clothing needs of the Bangladeshi living in London as much as the ultra-orthodox neighbourhood Mea Shearim functions in Jerusalem to serve the very particularistic religious needs of the ultra/orthodox Jews.

- *The café* is mostly associated with comfort in the city; it is an urban entity, which characterises many of the city's exciting sides. The café is what makes a place more public oriented – more for the use of the public rather than for the residents. The café has a different character in Jerusalem and London mainly because of different climate and cultural norms, but whatever its characteristics, its role in city life is similar and central in increasing people's sense of comfort.

- *Terrorism* has become one of the most obvious reasons for discomfort in the city, especially in Jerusalem because currently this city suffers the highest terrorist attacks in Israel. The fact that terrorism is targeted to destruct everyday life makes its effects on city life more explicit and dramatic.

'*The right to comfort*' consists of all the above elements. The right to comfort is linked with many aspects of human rights and citizenship because of its association with power relations and hegemony. The right to comfort also has its spatial–physical dimension with regard to politics of planning and development, which are implemented in the city. It concerns the level of infrastructure and services in the different parts of the city as indicating physical comfort. The right to comfort in this respect becomes again a part of citizen rights in the city, the right to obtain the required level of infrastructure and services from the municipality in exchange for the municipal taxes that its citizens pay.

Finally, the analysis of comfort at the different levels of space helps clarify the extent to which city people experience similar or different daily practices whether they live in London or Jerusalem. There are many shared universal experiences regarding comfort among the citizens of the two cities, both in the very private – the home – or other public spaces. Some of these have been presented in the concluding remarks above. However, since comfort is so much affected by power relations and control it should be noted that side by side with the similarities in daily experiences there are still major differences between the two cities. These differences come from the dissimilar political situations and the extent to which equality is penetrated in the two cities' politics of planning and development.

CHAPTER 6

On belonging

'My home, memory, ownership, family, friends . . . I don't belong anywhere else.'

This was Aziza's response to the question: 'what makes you feel you belong to your home?' Aziza defines herself as a Palestinian. She lives in West Jerusalem, has Israeli citizenship, although she does not identify herself as an Israeli. What are her reflections on her sense of belonging to the different environments identified in the book? This chapter aims to deconstruct this notion not only as an abstract element but also as a notion related to people's daily practices in the different layers of space, starting at 'home' and continuing to the 'city'.

In spite of the associations of 'belonging' and 'the holy city' with notions of sacredness, rituality, religious territoriality and conflicts, this chapter looks at the meanings of 'everyday belonging', the private secular sense of belonging. I am interested not only in understanding the rather obvious meanings of belonging to holy places, places of rituals, places appropriated for religious or nationalistic purposes, but in deconstructing the 'secular' aspects of belonging that are part and parcel of everyday activities in everyday spaces, activities which are ordinary for people's lives in the two cities.

Is there such a 'secular' sense of belonging and attachment to the different categories of environment (home, building, street, neighbourhood, city and city centre, urban parks) or can belonging be associated only with 'holy' places of rituals, symbolism and territorialism? Is there any personal sense of belonging or is belonging developed only within a communal context? These are some of the issues this chapter aims to deal with in analysing people's narratives in London and Jerusalem. These narratives represent the daily practices of people who are 'non-extremist' in their national, political, social and religious perceptions and everyday activities. The narratives shed some light on the meanings of this notion for the 'ordinary people', those who are in the majority in most cities. Obviously this selection of interviewees dictates and affects the content of analysis but it serves the purpose of highlighting the 'everyday life' and the 'secular' aspects of belonging.

Let us return to Aziza's narrative. Aziza actually related to the multi-dimensional meanings of belonging. As a Palestinian, a citizen of Israel, she talks about her home and associates it with memory. It could be her personal home as the question related to her personal 'home' but perhaps she also related to a national communal belonging and memory of the Palestinians of their homes before 1948. Her personal sense of belonging here is associated with nationalism. The research on belonging and nationalism has gained attention in the literature especially with regard to the discourse around citizenship and human rights of indigenous people (Yuval Davis 2000, 2003; Read 2000). Belonging in this respect is linked to notions of participation and inclusion in the construction of citizenship identity and membership of one's own nation. The notions 'politics of belonging' and 'politics of recognition' are linked to such a rationalisation.

Aziza also mentions belonging as associated with ownership. In this context a sense of belonging to the physical environment can also be associated with an emotional attachment created between a person and a physical place that is based on the person's subjective meanings of that place. Belonging here is linked to its material and physical aspects. In addition, Aziza mentions a sense of belonging as connected to family and friends. Belonging here is linked to the people living at home more then to the physical home. Aziza finally says: '*I don't belong anywhere else*'. Indeed the notion of belonging is associated with the discussion on sense of place, that is, what makes a place a home and what are the meanings of the notion of belonging in cosmopolitan environments, issues that will be developed later in the chapter.

What is a sense of belonging? Probyn (1966 in Yuval Davis 2003) has emphasised the affective dimensions of belonging – not just of being but longing or yearning. The *Oxford Dictionary* defines belonging as composed of three meanings: first, to be a member of (club, household, grade, society, etc.); second, be resident or connected with; third, be rightly placed or classified or fit in a specific environment. These dimensions emphasise the membership component of belonging and its multi-layered dimension (Yuval Davis 2003). In many cases belonging is a feeling that consists of past and present experiences and memories and future ties connected to a place and which grows with time (Crang 1998; Fullilove 1996).

De Certeau's *The Practice of Everyday Life* (1984) contracts the notion of belonging as a sentiment, which is built up and grows with time out of everyday activities. De Certeau terms it 'a theory of territorialisation' through spatial tactics. In his work he draws the distinction between 'place' and 'space' as, somewhat confusingly, space being place made meaningful (Leach 2002), or in de Certeau's words: '*space is a practical place*. Thus, the street geometrically defined by urban planning is transformed into a space by walkers' (p. 117). For de Certeau, everyday activities in the city are part of a process of appropriation and territorialisation:

> The ordinary practitioners of the city live 'down below', below the thresholds at which visibility begins. They walk – an elementary form of this experience of the

city; they are walkers, *Wandersmänner*, whose bodies follow the thicks and thins of an urban 'text' they write without being able to read it. (p. 93)

This everyday act of walking in the city is what marks territorialisation and appropriation and the meanings given to a space. What de Certeau constructs is a model of how 'we make a sense of space through walking practices, and repeat those practices as a way of overcoming alienation' (Leach 2002: 284). De Certeau actually defines the process in which a sense of belonging is established, a process of transformation of a place, which becomes a space of accumulated attachment and sentiment by means of everyday practices. Belonging and attachment are built here on the base of accumulated knowledge, memory and intimate bodily experiences of everyday walking. A sense of belonging changes with time as these everyday experiences grow and their effects accumulate.

Belonging and memory

'Belonging is very complex, subtle, nothing concrete, often it is familiarity with, memory from childhood with palm trees, I feel I belong more [to childhood landscape] as it is more familiar to me . . . (Suna, 40s, single, Egyptian – (British)–Mediterranean, London, 29 July 1999)

'No sense of belonging to the house I live in . . . I have a deep sense of belonging to the town of my family where "the graves are" – there I feel rooted, a sense of belonging to where my mum lives . . . I ask myself "where do I belong?" and I don't know!' (Eleonore, 50s, single, British – white, London, 1 September 1999)

Memory is perhaps one of the most explicit expressions of a sense of belonging and a part of one's own identity. Memory is either 'real', that is, a personal memory of childhood's reminiscences or it is a 'symbolic memory' consisting of ancestral graves, as Eleonore's narrative emphasises.

Sandercock (1998a) connects between belonging and life in the cosmopolitan city. She mentions three elements in the city that are crucial for creating a feeling of belonging: memory, desire and spirit:

Memory, both individual and collective, is deeply important to us. It locates us as part of something bigger than our individual existences, perhaps makes us seem less insignificant, sometimes gives us at least partial answers to questions like: 'Who am I?' And 'Why am I like I am?' Memory locates us, as part of family history, as part of a tribe or community, as a part of city building and nation making. Loss of memory is, basically a loss of identity. (1998: 207)

Memory in fact creates and consists of a sense of belonging. It could be a short-term memory based on intimate knowledge, which builds up in everyday life practices, of daily use of the streets, the paths, the pavements and the city centre. This is a physical memory because it engages bodily experiences in using these spaces: walking, driving, cycling. It is also an identity-related memory as it engages experiences affected by one's own identities as a woman, gay, black, disabled, etc. Memory is also long term. It goes back to the past and consists of an accumulation of little events from the past, our childhood experiences, our personal readings and reflections on specific spaces, which are associated with significant events in our personal history. Such memories build up a sense of belonging to the places where the events took place. Thus Suna, who is an immigrant living in London, develops a sense of belonging and attachment to her childhood places where her past memories are a major part of her identity. This role of memory as part of one's own identity or as part of a collective identity of the community becomes more and more evident in people's narratives:

> 'What makes me belong to my home is the people I love and they love me, the objects I love, colours, things that are part of my past – experiences, part of my home in me.' (Susana, 30s, married with one child, Israeli – Jewish, Jerusalem, 13 July 2000)

> 'I feel I belong to the Old City of Jerusalem as it bring memories from school days and boarding school, we used to go there every week. It makes me feel connected, it brings memories of my school days, in front of the Orient house. I used this area a lot in my life.' (Saida, 30s, single, Palestinian – Muslim–Arab, Jerusalem, 30 December 2000)

For both Susana and Saida memory becomes part of their own identity but also part of their collective identity. For Saida in particular, who as a Palestinian currently lives under daily oppression and humiliation, memory of places she 'territorialised' in her everyday practices in childhood is almost the only possibility she has to feel attached and connected to the city! Because of the political situation in Jerusalem her past memories are her link to the city rather than her current, limited everyday practices.

For Stuart childhood memories and belonging are also two connected issues. He says he does not feel he belongs to the neighbourhood he lives in precisely because he did not grow up there and has no memories that link him to the past:

> 'What makes me feel I belong to my neighbourhood is a sense of familiarity. I don't feel I belong to the area because I didn't grow up there, I have no friends or family there. I haven't got memories that link me to the place. If I bought a house then I would feel a sense of belonging . . . at the moment I feel like a guest in the neighbourhood, so it's about ownership.' (Stuart, 30s, single, British – Ghanaian–Afro-Caribbean, London, 6 August 1999)

For Stuart a sense of belonging is connected either to childhood memories or home purchase which makes links to the area at present stronger for him. We can see how memory is now perceived as a universally accepted need: 'Memory and tradition are keys to beginning to understand Aboriginal attachment to land', writes Hillier, 'memory is embodied in identity' (1998: 218). In Aboriginal culture memory is ancestral based. It creates and is part of the attachment or belonging to land. Memory thus is a significant component in the construction of a sense of place and as such it is also power related (Cuttis 2001). In many cases power is exercised over memory so that those in power can dictate which memory is spatially respected and commemorated. Australia's cities are one such example, where only in the last two or three decades have Aboriginals' memory and attachment been recognised and commemorated as part of city development and expansion (The Opera House and the Botanical Gardens in Sydney are examples of integrating Aboriginal sites of commemoration as part of modernised sites). The conflict of power as one of the major elements of belonging is also very explicit in the Jewish and Palestinian struggles in Jerusalem. While Saida's narrative above engages her personal memories and personal territorialisation of places in the city, Aziza narrates 'the national Palestinian memory' and reflects the situation in the city before 1948 when some of the neighbourhoods in West Jerusalem were populated by Palestinians:

'I don't feel a sense of belonging to the neighbourhood . . . it is a place of sadness because the original residents of Talbia [a neighbourhood in Jerusalem] are not there any more. So, the neighbourhood is "my" place but not really "mine" because most of the people I would have liked to be there are not there any more. My dream is not to go back to what it used to be but that the city as a whole will return to what it used to be before 1948, a city that everybody lives in. It was then more of a city than it is now . . . I want them [the Palestinian residents of Jerusalem] to return but not to return to that time . . . places with no dreams are bad places and this is what happens to East Jerusalem, a city with no desire, no dreams, no memory.' (Aziza, 30s, single, Palestinian – Palestinian (citizen of Israel), Jerusalem, 7 August 2000)

Here memory is collective and imagined. Aziza did not live in the neighbourhood before 1948 but she bases her imagined memory on the collective identity of the Palestinians in their everyday spaces in Jerusalem pre-1948, when part of West Jerusalem was Palestinian. Spaces after 1948 have changed. Similar to Australian cities, West Jerusalem has become Jewish but with no indication of its Palestinian past. On the contrary the nationalistic powers, which dictate the city's development and expansion, make sure that its Jewish demographic superiority will be kept (see Chapter 4). This raises the question of whether the prospect of the Palestinians' sense of attachment to the West City will ever materialise.

From the discussion on belonging and memory we can see the strong bond that exists between the two sentiments: to re-member means in many cases to be-long. The repetitive nature of re-membering is perhaps the synonym of the daily repetitive practices that de Certeau mentions as creating a sense of space and attachment.

As we have seen so far, memory and remembering can be personal and real or collective and imagined. Memory can be short term, consisting of personal – usually childhood – events, which transform public spaces into intimate ones. Memory can also be long term, constructed and based on a collective identity and memory of the community. The latter can become the basis of a sense of belonging to 'sacred' sites, sites representing symbolic and mythical histories, sites of rituals and symbolisms, which are so significant for many places in Jerusalem.

Between comfort and belonging

The associations of comfort and belonging emphasise the differences between what is more physical – a sense of comfort, and what is more emotional – a sense of belonging. Are these intuitive differences reflected in people's own experiences?

For some people comfort and belonging are sometimes closely related, with no significant distinction between them. Perhaps this is because most of the people interviewed are people with no extreme political, national or religious agendas so that significant expressions of religious or national belonging and attachment were not raised. As seen later in the chapter, 'everyday' belonging is a sentiment that is built up and changes over time as much as a sense of comfort does. For some people comfort is easier to define because it is associated with physical perceptions of lifestyle and sometimes it is almost related to basic human needs, unlike belonging, which is much more complicated because it evolves emotional and psychological aspects. One of the participants in the research said that one cannot live without feeling comfort, comfort is basic but people live their lives without feeling a sense of belonging. Could that be? Let us look at how people themselves use the two concepts:

> 'I belong to my home because I live there, I've always felt comfortable there, I can't explain why – perhaps chemistry . . . I always felt comfortable at this house, it's familiar, I've made bits of it full of my things, all my family photos.' (Suna, 40s, single, Egyptian – Mediterranean, London, 29 July 1999)

> 'Comfort makes me feel I belong.' (Rebecca, 36, married, Israeli – Ashkenazi, Jerusalem, 3 February 2000)

> 'The people in the house (make me feel I belong) . . . feeling of comfort, mental comfort and physical comfort – a home with the full meaning of the word. The place I feel most comfortable . . . a place of retreat.' (Neta, 50s, married with three children, Israeli – Ashkenazi, Jerusalem, 11 June 2000)

Here are several expressions of belonging which built in themselves the notion of comfort, either in Suna's and Rebecca's straight connections that comfort makes them feel they belong or in Neta's reflections of the connection between her sense of comfort and belonging in relation to the people in her home – her family. Here a sense of belonging is associated with 'being a member of' or 'be-longing'. A member of an emotional space – the home.

For Fatma the connection between the two elements is straightforward: 'Where you feel comfort – you belong':

> *'Where I feel comfortable I belong, if I feel comfortable in France I belong there, if I feel comfortable in the US I belong there.'* (Fatma, 40s, single, Arab – Muslim (Israeli citizen), Jerusalem, 18 April 2000)

As seen in Chapter 5, Fatma does not feel comfortable in her own home because she lives with her mother in a house in which she is not free to live her own life. For her belonging is definitely not connected with the home because her home is a symbol of cultural control, whereas in places outside her cultural context she feels free and therefore belongs. For those who are oppressed in their home environment a sense of belonging is associated with being anonymous and being free of restricted rules and norms, and this comfort actually allows a sense of belonging to develop.

Perhaps the clearest links between comfort and belonging are expressed in Magda's definition:

> *'For me comfort and belonging are connected. There is a physical comfort and there is a mental, psychological comfort and then it's belonging.'* (Magda, 30s, single, Palestinian – citizen of Israel, Jerusalem, 23 May 2001)

Magda's embodied knowledge of belonging is associated with that of comfort. Psychological comfort is belonging for her. Here the connection between the embodied knowledge and identities becomes clear. As a Palestinian living in Jerusalem 'psychological comfort' has a deep meaning associated with notions of equality, freedom and independency, which she mentioned in other parts of her narrative.

So far we have seen how the two terms are closely associated so that in many cases people of different identities use the two terms interchangeably to expose the deeper meanings of each. This reveals some surprising aspects of a sense of belonging, related for example to anonymity, which must be understood in a specific identity context of suppression, marginality and discomfort in one's own personal home context.

A sense of belonging and a sense of place

Is sense of belonging to home and city meaningful and relevant any more? In particular, is it meaningful for those who live in globalised urban spaces, where the role and influence of the nation–state is declining?

A sense of belonging both to people and places such as home is contested in the literature on globalisation. Several researchers relate the danger of living in globalised urban spaces where a sense of belonging is becoming less and less part of daily life. Others suggest different meanings to the notion of belonging, home and sense of place. In fact, the 'déjà vu' of these notions to academic and planning discourse, especially in the context of 'time–space compression', is probably one of the signs of the worries surrounding life in the global city. Massey (1994) describes these worries:

> Moreover, it is argued that this new round of time–space compression has pro-duced a feeling of disorientation, a sense of the fragmentation of local cultures and a loss, in its deepest meaning, of a sense of place. The local high street is invaded by cultures and capitals from the world over, few areas remain where the majority of industry is locally owned; places seem to become more and more similar and yet lacking the internal coherence; home-grown specificity is invaded – it seems that you can sense a simultaneous presence of every where in the place where you are standing. (p. 162)

Massey is one of those who challenge the validity of such descriptions. She doubts the extent to which these reflections really represent the everyday life of the majority of people in global cities, a doubt that we examine in narratives later in the chapter. Massey also doubts the existence of placelessness, which has become a major argument in the literature on postmodernity (see for example Harvey 1989). Places are identity related she says, but who identifies their iden-tity? In this context Massey mentions the connection between identity construc-tion of a place and the notion of nostalgia, that is, the memory of that place:

> In some sense or another most places have been 'meeting places'; even their 'orig-inal inhabitants' usually came from somewhere else. This does not mean that the past is irrelevant to the identity of place. It simply means that there is no internally produced, essential past. The identity of a place . . . is always and continuously being produced. Instead of looking back with nostalgia to some identity of place which it is assumed already exists, the past has to be constructed. (1994: 171)

Here Massey actually suggests that memory, which is the base of identity construction of a place and belonging, is not a one-dimensional process. It is rather changing and reproduced continually. In this context Massey quotes hooks, who warns that we have to distinguish between: 'a politicisation of memory that distinguishes nostalgia, that longing for something to be as once it was, a kind of useless act, from that remembering that serves to illuminate and transform the present' (Massey 1994: 171). Massey in fact challenges the almost obvious connections presented earlier in the chapter between memory, belonging and a sense of place. For example, she refers to the multiple mean-ings of a very particular place – 'home' – and argues that the home should be conceptualised both as singular and as bounded. Places can be home, argues Massey: 'but they do not have to be thought of in that way, nor do they have to be places of nostalgia. You may, indeed, have many of them' (p. 172). This reflects a period in which all places are interchangeable: 'where I put my hat down, that's home' (drover George Dutton in Casey 1993). Does this mean that the whole notion of belonging to home can be reshaped and become more geographically placeless then before? Home, according to this vision, is not necessarily a geographic location but an emotional attachment that can change from time to time and from place to place. This line of thought is also emphasized in hooks' analysis of home:

> home is no longer just one place. It is locations. Home is that place which enables and promotes varied and ever-changing perspectives, a place where one discovers new ways of seeing reality, frontiers of difference. One can't accept dispersal and fragmentation as part of the constructions of a new world order that reveals more fully where we are, who we can become. (hooks, 1991: 148)

Massey and hooks in fact deconstruct the notion of home and belonging not only to their physical and geographic associations but also to their symbolic or imagined connotations. A sense of belonging to home becomes a psychological and emotional sentiment and less related to specific physical locations. This sense of place is not less meaningful in the global era, it just has different meanings and contextualisations. Massey also provides a feminist perspective to this dialectic notion of home and belonging. The construction of 'home' as a woman's place is associated with notions of stability, reliability and authenticity. But it is also associated with power relations, as reflected in many of the women's narratives.

So far we have learned that a sense of belonging, home, memory and sense of place are all identity constructed and that these constructions are changing over time in an ongoing process that sometimes loses its geographical context. We have also learned that a sense of place associated with belonging and attachment is not necessarily irrelevant or missing in global urban spaces, as some researchers hint in their 'time–space compression' description. On the contrary, it is a sentiment that becomes even more significant and central in daily practices in current urban spaces, perhaps with different meanings and contextualisations. The next section explores this centrality by focusing on the extent to which the practices of everyday life that de Certeau analyses do generate a sense of belonging in the two cities, a sense of belonging that is not necessarily sacred and spiritual but more related to 'everyday or casual sense of belonging'.

Belonging and the notion of home

'What makes you feel belong to your home?' was the question I asked the people I interviewed. When we read some of the narratives we discover some similarities between Massey's and hooks' interpretations and people's reflections of home. For example, one of the dominant elements of belonging and memory at home was 'my personal possessions' and 'the order of my possessions'. Possessions as an expression of belonging reflect Massey's and hooks' interpretation of home as it means that home can be anywhere because possessions are mobile. Belonging to home is also about the people who live at home – family members. Those are 'mobile' too so that this is another expression of a placeless sense of belonging to home that is similar to what Massey and hook suggested. The dominance of possessions in one's sense of belonging at home is somehow similar to people's definition of what is comfort at home (see Chapter 5) but this point is emphasised more when related to a sense of belonging and has repeated itself many times in the narratives. Donald and Abraham express similar sentiments on the notion of belonging at home:

'My home is part and parcel of my relationships. It is part of me. I am very happy there, it's my heaven as well as my home.' (Donald, 50s, married, British – English, London, 5 August 2001)

'It is a mutual feeling. I belong to the house and the house belongs to me. I can do what I want in it, I can design it as I wish . . . in a sense the house is an extension of myself as much as I am an extension of the house.' (Abraham, 50s, married with three children, Jewish – Israeli, 29 February 2001)

These are two emotional expressions of two men towards their home which emphasise that 'home' is not only a woman's place. Its associations with stability, security and support are reflected in men's narratives too. It seems that belonging is about the freedom to be oneself, the freedom to live as one wants that is common to men and women although it is sometimes harder for women to achieve than for men. One of the expressions of such freedom is the ability or licence to arrange the home space in the way one wants and to arrange the possessions in one's own way. The order of possessions at home is actually 'the planning of the home space', where people express their own personality. Apparently the way possessions are arranged and their order in one's home symbolise belonging. Order of possessions is a reflection of power relations too, the power to control and to have the ability to influence how and where things are arranged, the power to territorialise. Here are some narratives that emphasise this point:

'I don't feel I belong here because the house is not arranged as I want, until the things in the house will be to my own taste and in their right place I will not feel I belong.' (Aliza, 38, single, Israeli (Jewish), Jerusalem, 23 April 2000)

'What makes me feel I belong to my home is the ability to have piles of stuff like books or magazines and I know exactly where they are and why they are there . . . other people may think it's a mess . . . I put them there deliberately! I am marking my territory.' (Robert, 30s, single, British – white, London, 1 September 1999)

Both Aliza and Robert associate the notion of *belonging with the power to control, to arrange the home space* according to their own needs. Here belonging is about owning, it is about the territorialisation and appropriation of space so that the power to control and to arrange it is achieved. This same link is reflected in John's association with a sense of belonging to his home, the fact that he made the kitchen makes him feel he belongs:

> '*What makes me feel I belong to my home is a sense of familiarity, accumulated investments and efforts, I made the kitchen and the garden. It is about personal investment.*' (John, 60s, British – white–Anglo-Saxon, London, 11 August 1999)

Here again the power to change, to arrange, to make a mess, to decide how space is shaped, is associated with a sense of belonging, of owning. This notion of the power and the ability to control the house order and the arrangements of inner spaces according to one's own needs is very strongly associated with belonging.

We saw in Chapter 5 that flat owning has been linked to the notion of comfort at home. It is associated with the notion of belonging as well, especially for lower-income people. Ronit is in her 40s, divorced and a mother of three children; she identifies herself as Israeli–Mizrachi. As a lower-income earner and single parent she had difficult times in moving flats. She says she feels very secure and more attached to her flat since she purchased it.

James expresses the same feelings of security and belonging as the owner of his house:

> '*What makes me feel a sense of belonging to my home? It's a contradiction in terms. If it was not mine I wouldn't feel I belong . . . What I just said is that it's mine, legally, culturally . . . personally I am there because I want to be there, my choices are mine.*' (James, 60, single, British–Jewish, London, 22 August 1999)

Nevertheless, flat owning is not always a condition of a feeling of belonging to home. Magda for example, who is 38, single, Palestinian (a citizen of Israel) feels very much that she belongs at her flat although it is rented. It therefore seems that flat ownership as an expression of belonging is associated with economic ability – for lower-income people it is more important to own a flat. Magda is a middle-income person and she lives in a flat owned by a Palestinian and therefore it is not essential for her to connect her sense of belonging to purchasing her flat.

A sense of *belonging at the home level* is sometimes connected to *making choices*. James mentioned choices in his narrative above regarding a sense of belonging at home. Judith also relates to making choices when talking about the meanings of belonging to her home:

> '*What makes me feel I belong to my home? Everything! I choose everything! Choice! That's what makes me feel I belong to my home.*' (Judith, 30s, married with two children, Israeli – Swiss, English–Jewish, London, 2 September 1999)

In general it seems that the more choice people have the stronger their sense of belonging becomes. Choice is in fact another version of power relations and control. The higher the level of control and opportunities one has, the more possibilities of making choices exist for a person. Obviously the smaller the scale the more choices people can make. For example, people usually choose the area they wish to live in and to belong to when they purchase a house or a flat. This association of belonging and choice or 'a belonging by choice' depends on people's financial abilities. Linda made the choice to live in a specific flat and thus she feels very much 'belonged' to it. She is 38, single, Israeli–'nomad' (as she defines herself), and has lived in London for the last 11 years:

'The home is my choice. I choose it to be my home.' (London, 30 July 1999)

Making choices and a sense of belonging are sometimes conflicting elements, as reflected in Muhammad's narrative. For him the act of choice as a construct of belonging is very ambivalent. He is in his mid 30s, married with one child; he was reluctant to define his identity and was only willing to say that he is 'cosmopolitan'. Muhammad is Palestinian – a citizen of Israel who has lived for several years in Berlin:

'I have returned from abroad and renovated the house, so that I feel at home. I painted it in different colours . . . then I understood that colours would not make a feeling of home. I was for many years outside home and I've changed in different ways from my friends . . . today, after six years I can see this house as my home, that I belong to it. The time factor plays a role . . . and it is not only the home, the home is like a cell in the city. In my house in Berlin I still feel at home . . . it's not only the house, it's also friends, people that are equal to you and you are equal to them . . . for the first three years [after my return] I felt like a stranger here. Why did I return? I believe that I can contribute to the society, the Arab society, and also to be here with the family, but it was a mistake, there is a gap between the world I lived in in Berlin and the world here, here it is like a third world and there it is Berlin.' (Jerusalem, 23 April 2000)

These are reflections of a person who chose to cross boundaries. In many ways Muhammad is like an immigrant in his homeland. This is because he lived outside his home for a relatively long time and found it hard to readjust. Muhammad also finds it hard to live as a Palestinian in Jerusalem where his identity means oppression and discrimination as opposed to his sense of freedom in Berlin. Home is symbolically used here to express the difficult situation of his own society in the city as a whole.

The articulation of 'belonging at home' for the Bangladeshi immigrants living in London sheds another light on the different meanings of *a sense of belonging for immigrants and a sense of belonging for indigenous people*. Belonging at home is a complicated issue for them to define. As immigrants they perceive 'home' not as a house or a flat but more as homeland; home is perceived as an emotional place, not a physical one. Those who were born in

Bangladesh and came to London during the 1970s as children find it hard to relate to the notion of belonging at home in its narrow meaning: '*Home is home there because we are here for economic reasons*' says Harun, who is in his 40s, married with five children and has lived in London for the last 38 years (London, 7 August 2001). He defines his identity as British – Bengali–Muslim. He did not choose to live in London; he came as a child and stayed because of what he termed 'economic reasons' and also because his children were born and live in London and will not go back to Bangladesh. He represents a generation of Bangladeshi who feel trapped between two worlds. They did not choose to move to London and now they cannot go back:

> '*We are trapped in a time zone. We can't go forward or backward*' (London, 7 August 2001)

Both Harun's emotional, 'imagined' home back in Bangladesh and his 'real' home in London are places where he does not feel he belongs. He cannot live in Bangladesh any more but he does not feel 'at home' in London. He feels neither British nor Bangladeshi. He is 'in-between-homes'. A sense of belonging for an immigrant is therefore an ambivalent issue and as such it is different from the sense of belonging of indigenous people such as Hassan. He is a Palestinian who lives in the Old City of Jerusalem in a house, which has belonged to his family for the last 600 years. Hassan is in his 40s, married with four children. He says: '*the house is part of me – the house is me*' (Jerusalem, 3 May 2001). For Hassan a sense of belonging to the house is deeply rooted in the chain-history of his family. He says he will never sell the house although it is too small for the whole family because it is part of him, of his identity. Two identities, immigrant identity and indigenous identity, which construct two forms of belonging: the dialectic sense of belonging of an immigrant who is also a member of a cultural minority and in contrast the strong sense of belonging of an indigenous person who expresses a strong bonding to their home and probably their city. The two men feel the same sense of belonging to their homeland and home, the difference is that the Bangladeshi left his home and the Palestinian stayed.

Finally, and perhaps on a different scale, a sense of belonging is associated with food and spices, which are culturally constructs and identity related, especially among immigrants:

> '*And the spices that I usually eat and cook with . . . Indian spices that you can get everywhere . . . I learned the names of the spices in English only five years ago when I arrived in London because at home [Canada] I would shop in Indian shops – food makes me feel I belong.*' (Mandy, 28, single, Canadian – Indian, Jerusalem, 16 June 2000)

One of the first things that immigrants establish when they arrive in a new country are ethnic food stores that sell specific food and spices. Sandercock

(2000) calls it: 'market mechanisms', which also include services such as lawyers, tax accountants, shoe repairers, etc. London is famous for its large variety of food stores and restaurants, which provide almost every type of food that exists in the world. This is one of the significances of the 'ethno-towns' such as Chinatown and Banglatown – the large variety of restaurants and supermarkets established in these areas that sell food and spices from their places of origin. Jerusalem too has a large selection of ethnic restaurants and shops stocking particular spices. The Old City markets are known for their diversity of spices, especially those associated with Oriental cuisine. When Mandy was interviewed she lived in Jerusalem and enjoyed the wide range of spices suitable for her own cultural food – Indian cuisine.

To conclude this section, the narratives on belonging to home shed some light on the diversified and various meanings of this concept. They clarify and highlight the strong connections between belonging and the ability and power to control one's own space and its arrangement and order. These connections are well known in the literature (Yuval Davis 2003) with regard to individuals and communities but it is interesting to find out that forms of belonging are power relation related at the home level as well. This point is elaborated with relation to gender identities in Chapter 8. The narratives also clarify the relevance of making choices in the foundation of a sense of belonging and finally they help to distinguish between the sense of belonging of an immigrant (especially of a cultural minority) that is much more ambivalent then the sense of belonging of indigenous people. So far the narratives show that the notion of 'belonging at home' have similar meanings for people living in London – the global city and Jerusalem – the holy city. They do not seem to be affected by either the globality or the holiness of either city. The only group that feels disoriented and fragmented seems to be the ethnic Bangladeshi immigrants in London.

Claim, the boundaries of belonging and notions of exclusion in public spaces

Can claim to a space be perceived as a form of belonging? A claim over 'public' space is one of the expressions of belonging in everyday life. A claim is usually 'informal', taking place as part of casual daily encounters between people or groups. It usually takes place when individuals wish to appropriate sections of public settings for various reasons, sometimes to achieve intimacy or anonymity or for social gatherings, which are mostly temporary. Claim and appropriation of space are a construct of the everyday walking practices that de Certeau (1984) notes. These practices, which are repetitive, connect between what is termed by Bell (1999) as performativity and belonging. Performativity means a replication and repetition of certain practices and is associated with ritualistic repetitions with which communities colonise various territories. These performances are acted in certain spaces and through them a certain attachment and belonging to place is developed (Leach 2002).

Claim and appropriation of public spaces is usually engaged with class power relations. 'Belonging through possession' (Lee 1972 in Carr *et al.* 1992) means that class-based identity determines the use of open space. Middle-income groups rely on formalised rules of ownership and view spaces that are not owned by private parties as belonging to everyone and not available for appropriation by any single group. In contrast, argues Lee, among lower-income groups the feeling of 'belonging' in any public space is based on the knowledge of the place and its inhabitants and it is more common to appropriate public space, especially in public parks, among lower-class groups. This rather elitist view shows that even notions of claim and possession can be used to produce biased and perhaps even prejudiced attitudes based on class issues.

Claim on a class base can take other forms of appropriation and territorialisation. These are the implicit but sometimes explicit rules of inclusion and exclusion that play a role in the structuring of society and space in a way which some find oppressive and others appealing. One such example are the big shopping malls in many cities all over the world that are meant to serve the needs of certain groups in society while making efforts to exclude others (youths, the poor, lower class, blacks, immigrants, etc.) and by that introducing a lesser or a greater degree of surveillance. Parks in big cities are another example of public spaces where inclusion/exclusion takes place (Zukin 1997). These spaces are usually guarded and watched with the intention of excluding those who are not following the rules.

'The politics of belonging form norms of inclusion and/or exclusion that result out of boundary formation which differentiates between those who belong and those who do not, determine and colour the meaning of the particular belonging' (Yuval Davis 2003: 3). The 'boundaries of belonging' are usually symbolic and may change according to the needs and goals of the hegemony. The power to exclude, which is based on 'the boundaries of belonging', becomes in many cases the power of planning, of monopolising space through zoning and the relegation of weaker groups in society to less desirable and attractive spaces (see Chapter 3).

Do the narratives of belonging as related to public spaces refer to any of the above notions of 'everyday walking practices', 'performativity and belonging', 'senses of exclusion and inclusion'? In what follows we look at how people in the two cities narrate their sense of belonging as related to the different categories of 'public' spaces.

A sense of belonging to your building, street and neighbourhood

The notion of belonging to one's own building is less clear and explicit in the narratives than the notion of comfort. People did not identify with feelings of belonging to one's own building, perhaps because of its ambivalent character of 'a space in between' the public and the private, that is, a space which contains one's own flat – the home – but also other flats of people who are sometimes strangers but living in the same building. These ambiguous feelings of belonging to the building are probably a result of a lack of choice when it comes to neighbours. People choose their home or flat and they usually choose people to share their home with, but they do not choose their neighbours. This lack of choice seems to reinforce the ambivalent attitude of the notion of belonging to the building and eventually it becomes a technical issue involving mostly building maintenance.

Belonging to the street and the neighbourhood seems to be much more meaningful. A sense of belonging at this level is associated with some of the typologies that were identified before. Notions such as nostalgia and memory, walking practices, exclusion and homogeneity characterise some of the experiences of belonging in the street and neighbourhood.

Nostalgia and memory seem to be important triggers for people's identification with the street and the neighbourhood they live in. The 'one piece' neighbourhood that James mentions in his narrative below is for him a reference of his attachment and belonging. This homogeneous nature is gone and the neighbourhood is becoming a transit place where he does not feel he belongs any more. The sense of familiarity, which is so much associated with belonging, does not exist any more for him:

'When I grew up the people in the street were one piece, a classical community of middle-class Jewish community. That has changed, now it's a sort of a transit camp, people move in and out, it's the same kind but they don't belong because they are not there for very long.' (James, 60s, single, British – Jewish, London, 22 August 1999)

Robert reflects his sense of belonging to the neighbourhood in the fact that shopkeepers know him:

'What makes me feel I belong to my neighbourhood is that people in the launderette and grocery and in the newsagents know me. I feel a loyalty to my area. If my area was in the news I'd be interested and also defend it.' (Robert, 30s, single, British – white, London, 1 September 1999)

This desire and longing for social contacts seems universal. Nevertheless the desire for neighbourhood homogeneity seems to be dominant in people's identification with these spaces in Jerusalem. The neighbourhood in Jerusalem is associated for many people with the 'community', a group of people with similar identities, interests and codes of culture:

> '*A neighbourhood is a sociological unit, it is a social–ethnic representation. My neighbourhood is quite elitist, one of the best neighbourhoods: not many old people, not a lot of noise from neighbours. In Jerusalem there is a meaning to the term "neighbourhood". In Rehavia there are a lot of Philippinos [labour migrants] who take care of old people, in Katamon there is a different type of population, Jerusalem is very segregated . . . in Baqa there are rich people but also public housing . . . my neighbourhood is very elitist, it used to be a neighbourhood of rich Arabs . . . the best neighbours are the religious [not ultra-orthodox]. The Sabbath is then quiet, no noise, the youth behaves nicely . . . and ultra-orthodox don't come to live in the neighbourhood as they wouldn't want to live with the less religious people.*' (Joseph, 40s, single, Jewish – Israeli, Jerusalem, 4 December 1999)

Joseph actually maps the 'boundaries of belonging' in Jerusalem's neighbourhoods. He gives a personal perspective of social and cultural segregation in his city and neighbourhood. He mentions two types of segregation on a national basis, between Jewish and Palestinians, and on a religious basis between secular and ultra-orthodox Jews.

> '*It is important for me that there are no ultra-orthodox [religious Jewish] people in the street . . . there is something scary about them, in my sub-conscious I am always bothered to what extent I should be considerate, to what extent there is peace and security . . . it's better without them . . . it is a free street so I feel I belong.*' (Jacob, 30s, married with one child, Jewish – Israeli–Jerusalemite, Jerusalem, 17 June 2001)

Jacobs's narrative can be read as an expression of fear of difference (Sandercock 2000), a fear of change, of the changing face of the neighbourhood or of new neighbours who belong to a different cultural grouping. In Jerusalem, this desire for secular homogeneity or the fear of religious appropriation of the street and neighbourhood is based on the experiences in some neighbourhoods of a process of 'intrusion and settlement' of the ultra-orthodox to secular neighbourhoods. This process means a slow appropriation of the street and the neighbourhood by way of the purchasing flats by the ultra-orthodox and slowly building up an extremely religious community. When they become the majority in the street or the neighbourhood they use their power to dictate norms of behaviour in public spaces such as prohibiting driving on the Sabbath, prohibiting the use of TVs or radios on the Sabbath and at later stages molestation of women because of their 'indecent' clothing. Jacob fears this process and therefore for him a feeling of belonging means that the street and neighbourhood will remain 'free' of ultra-orthodox people. A sense of belonging here is associated with the desire to maintain social and religious homogeneity, which means segregation and exclusion of those different from us.

What do ultra-orthodox people feel about these desires for exclusion? It reflects in many ways their needs for their own segregation and secluded life and their desires to maintain their unique lifestyle. Hanna was secular but converted to an ultra-orthodox belief and lifestyle. She is in her 30s; she got married six months before the interview took place. She is not yet fully ultra-orthodox but in the process of becoming so. Because of the dramatic changes in her lifestyle her boundaries of belonging are changing too. Hanna and her new husband had to choose where to live after they got married. Eventually they chose not to live in an ultra-orthodox neighbourhood but in a mixed neighbourhood of both religious (not ultra-orthodox) and secular people. Hanna explains why:

> 'There is some sense of belonging in living in Ramot [a neighbourhood in Jerusalem] because it is not an ultra-orthodox neighbourhood . . . as someone who is in the process of "returning to religion" (baalat tshuva) I would feel very constrained living in an ultra-orthodox neighbourhood with the style I am dressed in and the fact that I drive a car . . . it is very tense there and here it is more amorphous from a social point of view . . . because we need our time and space to convert slowly. When I have children I will definitely move to a more religious [ultra-orthodox] neighbourhood.' (Jerusalem, 26 June 2001)

This is another representation of the very deep need for neighbourhood homogeneity. Hanna represents the ultra-orthodox community that Jacob fears. But she actually expresses the same desire for seclusion, a desire to live in a neighbourhood with people similar to her, which is probably a common desire for people who live in both London and Jerusalem.

What is perhaps associated with this desire for homogeneity, both in secular and ultra-orthodox neighbourhoods in Jerusalem, is reflected in Banglatown in London – a neighbourhood that mirrors the cultural sense of belonging of its residents. How do the Bangladeshi people feel towards their neighbourhood? We have seen that they expressed ambivalent feelings towards feelings of belonging to their home in London and the question is whether it is the same towards their neighbourhood. It seems that the neighbourhood reflects a sense of belonging more than home. The Bangladeshi community in the East End of London have formulated a community or 'networks of belonging' similar to what they had back home in the Sylhet district in Bangladesh (Forman 1989). The social composition of the residents of the Brick Lane–Spitalfields area is made up of people from nearby villages from the same district in Bangladesh. This is probably why Ahmed, who has lived in the area for the last 30, years says:

> 'Yes, I feel I belong to the Brick Lane area but sometimes I feel that this is actually part of Bangladesh. I live in this area and the family who live below us is from my village, the other family is from the next village and so on.' (Ahmed, 40s, married with six children, Bangladeshi – Bangladeshi–British, London, 7 August 2001)

Figure 6.1   Street names in the Bengali language, Brick Lane, London

The Bangladeshi immigrants have formulated a social structure in the neighbourhood that keeps their old community ties and their imagined sense of
belonging to their homeland. In Chapter 4 I elaborated on how 'physical spaces
of belonging' have been created by the Bengali people in 'Banglatown', constructing spaces and services that serve the cultural and ethnic needs of the
community such as food stores, traditional clothing shops, street names in the
Bengali language (see Figure 6.1), music stores, mosques and travel agencies for
their frequent trips to Bangladesh (Figure 6.2). Banglatown has in many ways
become a representation of spatial belonging and difference. But unlike ultra-
orthodox neighbourhoods in Jerusalem, which can also be seen as a reflection
of belonging, the Bangladeshi in Banglatown do not exclude people of other
cultures and do not enforce their norms and traditions as much as the ultra-
orthodox people do in their neighbourhoods in Jerusalem (see the elaboration
in Chapter 8).

In both London and Jerusalem people's narratives reveal the connection
between daily walking practices and a sense of belonging. The knowledge of
the area reinforces a sense of belonging: '*I know the street, I live here, I know
the building – every stone of it . . . I know it more and more. A very intimate
knowledge.*' (Susana, 30s, married with one child, Jewish – Israeli, Jerusalem,
13 July 2000). People living in London have also mentioned this intimate
knowledge of the area, of its little alleys and short cuts: '*knowing the streets
makes me feel belonging*', '*knowing short cuts – it shows I know the neighbourhood*' (Robert, 30s, single, British – white, London, 1 September 1999).

Figure 6.2   The gate to Banglatown, East End, London

These rituals and walking practices reflect young mothers' daily practices as they walk a lot with the baby carriage and get to know their environment (see Chapter 8). People who walk their dogs several times a day mentioned this ritual as contributing to their sense of belonging to an area.

A sense of belonging and territorialisation is also expressed in different forms of possession. Several patterns of possession can be identified in Jerusalem, especially in low-income, disadvantaged neighbourhoods. Residents of these neighbourhoods 'appropriate' the pavement near their front door by surrounding this area with pots of plants and benches, marking a boundary between their appropriated 'private' space and the rest of the 'public' space (see Figure 6.3). The people expand the boundary between their 'private' home and what they appropriate into public/private space, that is, the entrance to their home, and thus they form an exclusionary pattern of possession or physical belonging that transforms the use of the area near their doorway to 'private' use, only for family members, and mark a boundary of forbidden space for pedestrians as it is becoming a part of their home.

Carr *et al.* (1992) notice that this form of appropriation is typical of low-income people who have no ability to own property or who live in houses that are too small. This could also be interpreted as a protest against their low level of living or an effort to improve their living conditions, similar to what Kallus and Law-Yone (2000) call the 'voices from below' – inhabitants' protests and objections expressed in daily actions to change the structure of public housing

Figure 6.3 Appropriation of the 'public' into the 'private' in Jerusalem

and by that to take a role in shaping and reshaping public spaces. In the neighbourhoods of Jerusalem such kind of appropriation functions to moderate the sharp contrast between the private, that is, the home or the doorway and the public, that is, the street. Such forms of appropriation, claims and belonging are culturally constructed. They are more typical of Mediterranean cultures such as Italy where 'people acted as if the streets were their homes' (Pitkin 1993). In Sermoneta, the street becomes an extension of the house:

> Chairs were readily brought from the inside and placed on the street or on the balcony overlooking it. Women spent time visiting with each other, sewing, knitting, talking, gossiping. Men usurped the space in front of the bars for card games or filled the piazza in the late afternoon, grouping in clusters for talk and bravura. (p. 98)

These practices show how these ambiguous boundaries between private and public spaces are not only class issues but also, and perhaps mainly, cultural issues. The descriptions Pitkin provides of Sermoneta and other Italian towns can easily be related to daily scenes in Jerusalem's neighbourhoods, but is less common in England. Both the weather conditions and cultural norms of privacy rarely allow these situations to occur in London.

It is interesting to look in this context at Kilian's arguments against the sharp distinction between private and public devices. He suggests approaching the two concepts as forms of power relations that exist simultaneously in a space: 'privacy is the power to exclude while publicity is the power to gain access' (Kilian 1998: 124). Do these definitions make sense for Mediterranean cultures? Apparently they make sense for every culture. Kilian argues that 'public and private spaces are meaningless terms in the absence of social inter-actions. To be considered "public", streets, squares and parks must operate under certain rules and exclusions that paradoxically limit their publicity' (p. 124). This means that privacy, the power to exclude, is a necessary part of all spaces along with publicity – the power to access. Those rules that exclude/include and create claim and belonging do make sense probably in all cultures, but their expressions are different in different cultural and political contexts: they become cultural informal rules in Italian towns; they become both informal and formal rules in Jerusalem, especially with regard to the Palestinians; and they become formal laws in some parts of London where roads are considered 'private' with no access to the public (see Figure 6.4). What were private estates in London are now becoming the 'gated community' lifestyle.

To conclude, the narratives on sense of belonging in the building, street and neighbourhood emphasise some of its exclusive nature. A sense of belong-ing in the street and neighbourhood means a desire for cultural, social and especially religious homogeneity. This is very explicit in Jerusalem, perhaps more so than in London. Power relations are again a major construct of belonging in the street and neighbourhood, reflecting people's abilities to appropriate and exclude.

Belonging to the city centre and the city

Do the narratives on belonging to the city and city centre[1] reflect different dimensions of belonging than other categories of space? Some of the narratives generated similar feelings, for instance the close connection between the spatial knowledge of the city and a sense of belonging. Here de Certeau's (1984) ideas of repetitive walking practices increasing a sense of belonging and attachment are reflected in some of the narratives. In this section however we concentrate on two dimensions of belonging that were mentioned when people related to

Figure 6.4   Private road – no parking in London

the city: the nature of belonging in cosmopolitan times both in London and Jerusalem and the links existing between belonging and nationalism for the Palestinians in Jerusalem.

Belonging to the global city

> '*I don't feel I belong to a place but to people. I feel I belong to Israel, it is my identity home, but I don't feel I belong to the landscape except of Jerusalem. I feel I belong to Geneva, to the city centre, because I know the shops and the people, I say "Hey" to people . . . I feel I belong to certain parts of the Swiss culture . . . I don't feel I belong to London but two weeks ago when I was out of the UK I found myself saying: "We, the English", so there is a sense of belonging when I am outside . . . that is why I feel I belong to people and not to a place.*' (Judith, 30s, Israeli – Swiss–English–Jewish, London, 2 September 1999)

Judith's narrative best expresses the complex meanings of belonging for people whose roots are in multiple places. Judith lives in a 'network of belonging' and her emotional ties are spread in different places. The way she defines herself also reflects those split identities that symbolise people's lives and social networks in globalised cities. Another perspective of life in the global city is reflected in Suna's narrative:

'I belong [to London] because it is a cosmopolitan city. I am a foreigner but there are other foreigners . . . tourists . . . I feel that there is a sense of tolerance. I feel much less comfortable in a village in the countryside where everybody is English–white . . . there I feel an outsider . . . I feel as if they have never seen an Egyptian, a Muslim, and they don't understand my English.' (Suna, 40s, single, Egyptian – Mediterranean (British citizen), London, 29 July 1999)

City people and country people perceive 'the Other' differently. The city's people are more open, welcoming and tolerant, and the anonymous atmosphere welcomes 'the Other', whereas the countryside's people are more exclusionary and alienating to 'the Other'. Suna's reflections remind us of the work of the black British photographer Ingrid Pollard, who in 1984 exhibited a display entitled 'Pastoral Interludes'. Her work and photographs visually emphasise the discourse around belonging (Mitchell 2000) by exhibiting photographs of a black woman (herself) out for a walk in the English countryside. The English countryside represents belonging and Englishness, which in fact becomes a form of exclusion. Each of these photographs is accompanied by a text: 'Feeling I don't belong. Walks through leafy glades with a baseball bat by my side' 'it's as if the black experience is only lived within an urban environment. I thought I liked the LAKE DISTRICT, where I wondered lonely as a Black face in the sea of white. A visit to the countryside is always accompanied by a feeling of unease, dread' (Mitchell 2000: 260–1).

What both Suna's narrative and Pollard's photos emphasise are the significant characteristics of the city, which allow people of multiple identities to feel they belong. Belonging here is associated with acceptance, tolerance and anonymity. Anonymity can be perceived as a contrast to belonging but perhaps this is what belonging in cosmopolitan times means, to be included as part of the crowd in spite of one's own different identity.

The cosmopolitan sense of inclusion in London, mentioned by Pollard and Suna, is also expressed in the way the young Bangladeshi generation perceive their sense of belonging to London. Gaziul was born in Bangladesh and came to London when he was eight years old. He is now in his 30s and married to a Bangladeshi wife, whom he met in London not in an arranged form of marriage. Gaziul has a business in the Brick Lane area. He feels a sense of belonging to London although he emphasises that when he says: 'back home' he means Bangladesh, but then he adds: 'Britain is home too':

'Where do I belong? In Britain I know the system but Britain is not like America. Here you have the English society and then us, but I do feel I belong. We discuss it among us, we have no choice – we can't be anywhere else, we'll never be English men as much as they are but we are British, we have a British education, British mannerisms, British thoughts.' (Gaziul, 30s, British – Asian–Bangladeshi, (British citizen), London, 7 August 2001)

Gaziul's narrative expresses so clearly the ambivalent attitude towards feelings of belonging that he and his generation share. As already emphasised in

Harun's narratives (belonging to home), the Bangladeshis have their split or mixed feelings of belonging. Gaziul feels he has no choice but to live in London because he knows the system, the manners and the codes of behaviour. He says he could not work in Bangladesh. At the same time he will never feel English or be accepted as an 'English man'. He expresses feelings of exclusion on a social basis not so much related to London as a city but to the sense of exclusion from English society as a whole. It is interesting to note how he distinguishes between 'English' and 'British'. As he says, he will never feel within the boundaries of Englishness but he feels included within the boundaries of belonging of the British identity. It seems that Gaziul perceives British citizen identity as more inclusive then English ethnic identity, which is perceived by him, Suna and to a certain extent Pollard as exclusionary.

This is the right place to mention Eade's (1997) research, which analyses the politics of cultural differences and the process of identity construction as a global trend among young educated Bangladeshi living in London. 'As national belongings in the west are challenged by local and more global imaginings of community', he writes, 'the assertions of a "British", "English", "French" or "Belgian" cultural hegemony become more tenuous – the rhetoric of national unity notwithstanding' (p. 147). According to Eade those young educated Bangladeshi have constructed 'ethnoscapes' – where they interpret their group identity. Here the notion of belonging is analysed as the intersections between national, regional and religious constructions, reflecting a range of possibilities of creating a sense of belonging more as a mixture of all of these levels:

> These young people's understanding of what it means to belong to a nation-state is clearly shaped by political debates and practices in their country of origin, especially those which establish an association between a national [Bengali] culture and Islam. At the same time they engage in national constructions of belonging to Britain and reflect on their experience as Bangladeshi, Bengali and Muslims in a nation-state where they constitute a minority. (p. 153)

As reflected in the narratives of the Bangladeshi and in Eade's analysis the notion of belonging is part of an identity construction in global urban spaces, which allows them to construct their Britishness and maintain their Bangladeshi ethnic identities.

Belonging and nationalism in the holy city

While in London minority people such as the Bangladeshi distinguish between a sense of official inclusion within British citizen identity and exclusion from English cultural identity, the Palestinians in Jerusalem do not make such a distinction. They feel excluded both on their ethnic/cultural and citizen identities, that is, they are discriminated on the basis of their citizen identity by not being provided with equal services and infrastructure and they are denied their different cultural and ethnic norms by not being supported financially as much as the ethnic minority Bangladeshi are in London (See Fenster 1996). In this section we elaborate on notions of belonging as connected to a sense of nationality of the Jewish and Palestinians living in Jerusalem.

The Jewish perspective of belonging is multi-dimensional. It is constructed by the person's political and ideological affiliation and experience and one's own reflections on Jewish collective memory and belonging. Joseph and Rachel's narratives represent different perspectives and rationalisations of belonging. Joseph represents the Jewish hegemonic perspective of belonging in Jerusalem, although he mentions in his narrative the many other historical sites and forms of belonging that exist in the city:

> 'Historical sites excite me a lot. All the archaeological ruins, the city walls, the historical sites . . . As a professional [historian] I am excited about them. Not only Jewish sites but also the historical religious variety of the city, there is nowhere in the world where the variety is so high in terms of religious sections of the monotheistic religions: the Ethiopian, the Protestant, the Catholic and the Provo Slavic Churches all in the same place. Also in Islam; the Islamic Variety is fantastic, people who arrived from many places to integrate here. It is a city of extremes: lovers of Jesus, Hebrew Union College, ultra-orthodox extremists and the pick of the secular lifestyle – discos.' (Joseph, 40s, Jewish – Israeli, Jerusalem, 4 December 1999)

Joseph's sense of belonging to the city, which is based on the Jewish hegemonic memory, also emphasises the links to historical bonds that each religion establishes in order to sustain its claim to the city. His sense of belonging is based on the past and unique historical role of the city. Joseph doesn't emphasise the conflict of power relations that characterises the city's current daily life, but highlights its historical mixture, which for him is a major part of his sense of belonging to the city.

Rachel represents a critical view of belonging, based on the resentments she feels towards current life and events in the city. It is a dialectic and ambivalent expression of belonging. She is European by birth but chose to live in Jerusalem and convert to Judaism. She says:

> 'What makes me belong to the city centre and the city? That I live here, it was my dream, to come to Israel and to Jerusalem. This is the centre of my life; it is the centre for many people in spite of all the disappointments from the city: aggression, vulgarity, provincial feelings, and the Mayor that I hate, and also the segregation between secular and religious people that is worrying. Religious identity is so strong here. This city has such potential to be the most wonderful place in the world and it hasn't happened. Very frustrating. People come from all over the world to Jerusalem and it is still provincial . . . it is not the same feeling as in a capital in Europe . . . here there is no feeling of a capital city.' (Rachel, 30s, Israeli – European, Jerusalem, 24 March 2000)

Rachel emphasises the centrality of Jerusalem's 'holiness'. She is critical about the way the city functions at present. It seems that for her the 'holy' city cannot become 'global' or, as she says, like any other capital in Europe because of the very strong and sometimes negative attitudes of the different religious parties, which affect the city's life. Rachel perhaps emphasises how the formal, official 'top-down' municipal and national policies, which were discussed in Chapter 4, have their affect on everyday experiences of the people living in the city. What prevents Jerusalem from becoming 'global' in her eyes is the manipulation of its 'holy' character for particularistic interests. Joseph and Rachel both feel a strong sentiment of belonging to the city, but each bases it on a different temporal context. Joseph associates his sense of belonging with the history of the city and Rachel feels a sense of belonging, but she is also bitter about the city's current politics.

An opportunity to understand the Jewish religious perspective on belonging to Jerusalem was afforded me when I met Hanna – she was introduced earlier in the chapter (see p. 170). In her narrative she expresses very clearly how Jerusalem's holiness is complicated:

> 'This city['s identity] in terms of ultra-orthodox influence is the strongest, thank God! When I was still secular it was difficult for me to deal with this city but now I know, Jerusalem is the holiest . . . it is the centre of the world . . . In Judaism there are three holy cities: Zfat which symbolises air, Tiberia which symbolises water and Jerusalem which symbolises stone. It is built on a mountain. It is a harsh place, a place of dinim [strict religious rules], a place of confrontation with one's own self. A lot of people don't like Jerusalem and they say it's because of the traffic . . . but basically the reason is different. It is because Jerusalem makes people confront their own personal issues.' (Jerusalem, 26 June 2001)

Jerusalem is perceived not just as a city but also as a place of spiritual experiences, where people are forced to confront their own emotional issues. Surprisingly, not only religious people express this somewhat spiritual attitude. For Jacob, who is secular, living in Jerusalem is perceived as an emotional experience too:

> 'I feel that Jerusalem helped me to build up my own identity. My way of thinking has been developed to what it is today only because I live in Jerusalem, the city enriched me with its cosmopolitan character, it allowed me to open up to a variety of cultures, to open up my mind . . . Nevertheless, it is a harsh city and it's difficult to live here.' (Jacob, 30s, married with one child, Jewish – Israeli, Jerusalemite, Jerusalem, 17 June 2001)

Belonging in the holy city is spiritual both for religious and secular people. Here Jacob perceives the city as 'cosmopolitan', perhaps because of its holiness, a combination that makes it a harsh place to live in.

Elizabeth's sense of belonging to the city as a physical entity is interesting because of her connections to both London where she was born and lived up to her early 20s and Jerusalem where she has lived for the last 30 years:

> '*Three years ago I thought to leave Jerusalem because of the ultra-orthodox people's expansion but I'm over it . . . I am astonished by the beauty of the city. Here and also in London there are beautiful buildings, beautiful trees, beautiful landscape: the stones . . . there are beautiful buildings and beautiful renovations in Jerusalem. In London it's about the old buildings that were built 200 years ago – the churches and cathedrals . . . the wealth in these buildings which is an indication of the stability of the country, it is rooted, nobody can destroy the country. Here there are no such buildings, everything is new, well, there are Ottoman buildings but there was no empire here like the British empire. In England you feel the strength that used to be, here everything is so fragile. In London I have faith in the city although the IRA also planted bombs, but here I feel that everything is so fragile all the time because of the political and the military reality . . . who knows? Every hundred years the area changes regime . . . I would care a lot if something happens now. This is for me belonging to my home.*' (Elizabeth, 50s, British – Israeli–Jewish, Jerusalem, 16 March 2000)

Elizabeth's sense of belonging is connected both to Jerusalem and London and actually to the differences between them. Her embodied knowledge of belonging is constructed from these two localities and her narrative of belonging is about the expressions of city life as manifested in building styles. But perhaps her sense of belonging is about the sense of political stability that she feels in London, which makes the fragility of Jerusalem even more explicit. Her narrative clearly demonstrates the connection between urbanism, city spirit, city life and the politics of power relations in nation–state structures that have their effects on people's daily practices in the city.

Some explicit connections between belonging and memory are expressed in Neta's narrative. For Neta a sense of belonging is connected to the preservation of old buildings, a point mentioned by many others:

> '*My heart is in pain for every old building that is knocked down and replaced by a new building. I love old buildings. There is a need to preserve the old, the old character of the city . . . and also the trees, every tree that "they" knock down makes me feel sick.*' (50s, Israeli – Ashkenazi, Jerusalem, 11 June 2000)

Neta refers not only to the preservation of historical buildings, which has become a common practice in many world cities as part of the 'old and new' approach (see Part IV), but to more considered planning decisions that involve demolishing old buildings for the construction of new ones. This is one of the known spatial expressions of modernist planning that was also reflected in the demolition of part of the Spitalfields Market for to make way for offices (Chapter 4).

The Palestinian perspective on belonging is very much loaded. Their narratives concern mainly expressions of discrimination, oppression and abuse of their rights:

'I have a problem with this city . . . it is hard for me to identify with it after the occupation [in 1967] and the land expropriations from the Palestinians in order to build Jewish neighbourhoods and in order to destroy the territorial continuity of the Palestinians . . . the inequality and injustice . . . This city is so rich with great elements . . . other countries would have wanted to have it . . . for example, the encounter between desert and greenery, snow and desert, the topography, its history . . . where is all of this? Fucking country . . . A Palestinian feels comfortable in Tel Aviv and feels shit in the Old City of Jerusalem!' (Muhammad, 30s, married with one child, Jerusalem, 23 April 2000)

Exclusion and alienation are common experiences in Palestinians lives in Jerusalem. In Chapter 4 we elaborated on how the official policies of the government of Israel and the Jerusalem Municipality towards its Palestinian citizens cause violation of human and citizen rights: the politics of Palestinian land confiscations for the purpose of building Jewish neighbourhoods and the deep discrimination and inequality in municipal services are just few examples of the violations of citizen rights. Muhammad's perception of the city clearly reflects this situation when he says that he feels more comfortable in Tel Aviv than in his own city. Here Tel Aviv is perceived similarly to London as 'global' and cosmopolitan as opposed to Jerusalem, which seems so rigid and harsh. Muhammad's narrative emphasises the role of the politics of planning as serving the interests of Israeli politics in making clear political statements by means of planning schemes, in this case the construction of Jewish neighbourhoods on confiscated Palestinian land.

Aziza expresses similar feelings, but as a Palestinian who lives in West Jerusalem her sense of locality, of belonging, is expressed in a utopian 'imagined united city space', as she explains:

'Public activities, cafés – this is what makes the city real . . . and as many new activities are opened up the better it is . . . to see a new café open up . . . it's like looking at a flower. There are not enough libraries in the city. I live close to the theatre and the cinemateque and to the sadness of East Jerusalem, the part in the city that I need as it's part of my culture and I don't want to disconnect myself from it, I want it to be a ['normal'] urban space. In East Jerusalem I want more of cultural activity, urban activity, theatre . . . Then it will be one urban centre. Urban spirit is suffocated by having two separate city centres (East and West) . . . this is my imagined city, that it becomes one city centre and that all institutions exist in one united space.' (30s, Palestinian – Palestinian (citizen of Israel), Jerusalem, 7 August 2000)

Here again a connection is made between elements of citizenship, that is, equality in services and infrastructure and city vibrant life. For Aziza, to have one united city centre with all the facilities is a prerequisite for both a 'healthy' city life and her own identity needs.

Finally, Mandy's narrative, which represents an experience of a non-Jewish foreigner who lives in Jerusalem. She is in her 20s and identifies herself as

Canadian – Indian–Muslim. She is of Pakistani origin but was born in Canada and lived and studied in London. At the time of the interview she lived in Jerusalem with her Jewish partner. She says:

> *'This is a community in uniform . . . I am not Jewish . . . I am different, I have a strong sense of belonging to the Old City – to the Dome of the Rock area [a holy place for Muslims] where only Muslims can go . . . I feel a sense of spirituality there, I thought I would live in constant tension because of the situation [the political conflict] but it's much less in real life . . . the main sense of belonging is from the religious part. It is also where the strong sense of dis-belonging comes from . . . I live in the Jewish part of the city. I feel more commonality with Jewish people then with Christians. On a spiritual level, belonging is here but my connections are to Pakistan, India and Bangladesh. In London there are so many people from these three countries . . . so I felt more that I belonged there, also because of the language, here I live because of my partner, in London I would stay anyway, here I also can't get a visa as a non-Jew.'* (Jerusalem, 16 June 2000)

Mandy represents a type of cosmopolitan, uprooted person, with multiple layers of identities and networks of belonging. It seems that the main sense of belonging derives from her religious identity. Mandy expresses her spiritual belonging, which is also her spatial dis-belonging. She feels attached to the Muslim religion but she lives in the West – the Jewish side of the city. In her narrative Mandy highlights the different forms of inclusion and exclusion inherent in principles of citizenship and residency. As a non-Jew she can not get a working visa and if she is not married to a Jew she can not live in the country as a citizen or resident. It seems that most people interviewed (mostly secular) perceive Jerusalem as a harsh city because of its 'holiness' and the way this holiness is manipulated by the politics of the city.

To conclude, the analysis of a sense of belonging to the city and city centre highlights its multi-layered character as expressed in both cities. People express feelings of exclusion, mostly those who would be defined by the hegemony as located outside the boundaries of belonging – the Bangladeshi in London and the Palestinians in Jerusalem – although the degree of exclusion and the feelings of alienation are much stronger among the Palestinians in Jerusalem. In London, different forms of belonging are constructed for immigrants, forms that include 'networks of belonging', which are associated with people more than places.

Belonging to urban parks

Groups with different needs, norms, and paces may use time rather than space as a means of separating activities that might collide or create conflict . . . sometimes areas are subject to time sharing, with different groups claiming the space and establishing their territories at different hours. A playground used by young

children and their parents during the day may become a teen hangout at night.
(Carr *et al.*, 1992: 165)

'Time sharing' is a very common pattern of use in most parks located in borderline areas between East and West Jerusalem. There, time sharing usually takes the form of the Jewish and Palestinians using the park space according to their own time constraints. For example, Jewish people use the park areas mostly on Saturdays while Palestinians use the park areas mostly on Fridays, which is their holiday. However, this trend of time sharing in parks was not part of the daily experience of the interviewees in either Jerusalem or London. For them belonging to urban parks is usually about the knowledge of its various parts, a result of their daily use. People like the greenery, the open spaces in the city, the relaxed atmosphere and the possibility of going for a walk in it. Some said that they feel a greater sense of comfort than belonging in urban parks. For Susana parks are an essential part of the city:

> *'Parks are part of my scenery, "mother earth", I feel very connected to nature and I need to be in the park; to watch the greenery, the flowers. Cities without parks are not real cities!'* (Susana, 30s, married with one child, Israeli – Jewish, Jerusalem, 13 July 2000)

Parks are perceived by people as the symbol of nature in the city and therefore their existence makes people's sense of belonging to the city stronger. The park is a public place, a place for all. This is what makes Suna feel she belongs. But she also says that sometimes she feels fear in parks because of other people's norms and perhaps notions of appropriation, and power relations in dictating norms of use of the park spaces:

> *'I feel that I belong as it is a public space so everybody belongs. I don't crowd anybody's space. In residential parks I feel threatened as its residents use it. I am worried about not knowing what to do or not knowing the behaviour that locals know.'* (Suna, 40s, single, Egyptian – Mediterranean, London, 29 July 1999)

Suna mentions cultural rules of behaviour that she needs to know so that she feels a sense of comfort and a sense of belonging in urban parks. Suna's narrative problematises the publicity of urban parks, that is, their use by different cultural groups, which temporarily appropriate certain areas in the park and cause discomfort and a sense of alienation to people of other cultures.

Summary and conclusions

This chapter focused on the everyday meanings and practices of belonging, meanings and expressions that are not necessarily ideological, connected to religion or nationalistic but rather meanings and expressions based on personal, private, intimate daily experiences. The chapter's aim has been to understand and conceptualise the different 'formations of belonging', both collective and personal, and in particular the role and functioning of a sense of place as associated with a sense of belonging. A sense of place is such a controversial notion in current literature on globalisation that it becomes important to discuss its multi-layered meanings in the different levels of environment. Let us conclude the various meanings of belonging as they were highlighted in people's narratives:

- *A sense of belonging is associated with memory.* For example, the place one was born in, the family one belongs to. Other aspects of belonging are changing – places of living, homes and neighbourhoods. Some components of belonging are short term, others are based on long-term memories, such as childhood memories. Belonging has its personal aspects: belonging to places and people that are connected to personal experiences, personal memories. A sense of belonging is also collective. It is based on collective memories and shared symbolism of a community. Its significance in one's own life is a result of one's own affiliations, beliefs and ideology.

- *Belonging and walking practices.* Repetitive daily walking practices are one of the mechanisms of creating an 'everyday' sense of belonging. We all belong because we all have repetitive daily use of city spaces either on foot, by car or public transport. Our daily practices help us to draw our 'private city' and to underline the intimate alleys and paths that we use in our daily lives in the city. For young mothers and dog owners daily repetitive practices create their sense of belonging to the city.

- *A sense of belonging in the neighbourhood is usually associated with the neighbourhood's social, ethnic and religious homogeneity.* In Jerusalem this character is another mechanism of exclusion, mainly on a religious base between the secular and ultra-orthodox or between Jewish and Palestinians. In London, neighbourhood's homogeneity is usually more economic and not religious or nationally oriented.

- *A sense of belonging is about power relations and control,* even in intimate and private spaces such as home. The power to arrange the home space is what makes one feels a sense of belonging. The more the home space is arranged so that it meets the needs of the individual the more belonging they feel. The larger the category of space the more significant is the role of power relations on one's own feelings of belonging. In public spaces power relations are identified as 'claim', 'appropriation', 'exclusion', 'discrimination'. Power relations also dictate 'the boundaries of belonging'. They are formed by the hegemony and exclude the 'Other', those who are not considered by the hegemony as part of it such as the Palestinians in Jerusalem and to a lesser extent the Bangladeshi in London. The latter feel excluded from the boundaries of Englishness but included within the 'boundaries of Britishness'. A sense of belonging and power relations are associated with the 'private' – the power to exclude and the 'public' – the power to gain access. *Making choices* is another expression of power relations, which has its effect on the sense of belonging that one feels to one's own environment. The power to make choices in life is the power to control one's own life. The more a person has the power to choose a place to live and the more choice they have to change it the higher their sense of belonging becomes.

- *Belonging as a form of citizenship* is one of the more common interpretations of this term. Official belonging is usually formalised in patterns of citizenship. In Jerusalem, forms of belonging and citizenship are connected to the abuse of human and citizen rights of the Palestinians living in the city, usually by means of the politics of planning and development, which promote Jewish interests.

- '*The right to belong*' can be identified as the right of people of different identities to be recognised and the right to take part in civil society in spite of one's own identity differences, what Sandercock (2000) terms 'the right to difference'. The right to belong in contested spaces can be perceived as a deeper expression of 'citizenship in the global city'. The right to belong relates to the situations where one's own rights to equality and one's own rights to maintain identity difference are fulfilled. These rights are connected to communities' privileges to maintain their sites of memory and commemoration and these sites are acknowledged and preserved by the politics of planning and development.

- *Belonging and urban planning.* If one of the challenges of urban planning and management in the era of globalisation is to make our cities feel like home, one of the ways of doing it is by incorporating elements that are associated with a sense of belonging at home into urban planning and design. People mentioned in their narratives the 'physical objects' in the environment that would reinforce a sense of belonging, such as community clubs, more green areas for relaxing, more local shops, more areas to walk, more bridges for people to cross the Thames, more homes in the city centre, shade in the streets, vistas and views and intimate places in the city. Another important connection between a sense of belonging at home and urban planning is the association of order and belonging. Deciding upon the order of things is actually one of the basic activities of physical planning in each level of space, and the more people are involved in decision making about 'the order of functions' in their own street, neighbourhood or even city centre the deeper the sense of belonging they develop to these environments. Indeed several alternative planning traditions include mutual learning processes (especially the collaborative planning tradition, see Chapter 2) but perhaps the connection between people's involvement in the planning process and increasing sense of belonging is the important element to acknowledge here.

Note

1 People related to notions of belonging to the city and to the city centre interchangeably. For some people the city centre is the city and for some the only meaningful
space is the city and not its centre. In Jerusalem especially there was a tendency to
reply to belonging to the city rather then the city centre, perhaps because the notion
of belonging to the city is more relevant.

CHAPTER 7

On commitment

'*Maintaining the home, doing a good repair and keeping the garden in order make me feel committed to my home.*' (Alice, 41, single, South African (white), London, 28 July 1999)

'*Commitment to my home means that I am here for my children, for my parents, brother, grandmother – that I am on the other side of the telephone line for them . . . it means that I will be there whenever they need me.*' (Judith, 30s, married with two children, Israeli – Swiss, English–Jewish, London, 2 September 1999)

'*Commitment means cleaning the house once a week.*' (Rebecca, 30s, married, Israeli – Ashkenazi, Jerusalem, 3 February 2000)

'*Commitment to city centre? Not much. I don't like pollution, I want to preserve old buildings.*' (Robert, 33, single, British – white, London, 1 September 1999)

'*Commitment to the city centre means making sure that the centre will function as a centre, as a mixed centre, that the city will function as a city, commitment to the city or city centre means getting organised so that "they" will not take the city from its centre, that there will be shops, a centre, pavements, cafés – a centre that will make the heart . . . pump.*' (Aziza, 30s, single, Palestinian – Palestinian (citizen of Israel), Jerusalem, 7 August 2000)

These narratives represent the wide range of associations people expressed as related to a sense of commitment. They range from cleaning duties and home maintenance to emotional commitments to family members or the preservation of old buildings as part of a sense of commitment to the city.

The *Oxford Dictionary* defines *commit* as: '1. To do, to perform, 2. To entrust for safekeeping or treatment, 3. To pledge, to bind with an obligation'. A sense of commitment usually refers to people or ideologies.

In what follows I argue that commitment is the driver, the motivation for people to act, to change or maintain elements and dimensions in their environment[1] that make them feel comfortable and make them feel they belong. To feel committed to act is to feel that you care about your environment. In this respect commitment to public matters is very much associated with citizenship identity and with citizen actions regarding planning and development.

Another argument developed in the chapter is that commitment to the environment is multi-layered. There are different levels of commitment and different meanings of commitment in the different levels, as observed in the narratives above. Commitment can be personal or public, passive or active, physically oriented or socially and ideologically oriented. It is a sentiment that can cause internal conflicts between one's commitment to oneself and one's commitment to other values or people. Commitment can cause external conflicts between groups or communities whose sense of commitment clashes or between communities and authorities that sometimes abuse communities' rights in the process of development.

Crang (1998) says that involvement and commitment, like belonging, is subjective and is related to certain stages in the life cycle and identities. For example, a sense of commitment as parents is in general a much clearer notion than a sense of commitment for people with no children.

As seen later in the chapter commitment is not always a clear and easy sentiment to define, especially when related to the environment. For some people this notion has been identified as meaningless or meaningful only in relation to certain spaces such as 'the home' and 'the city'. It is nonetheless important to disentangle this element with regard to people's everyday experiences in the city because it is becoming a significant component of civil life in the age of globalisation.

To understand this sentiment the chapter analyses the different meanings and perspectives of commitment as they are associated with concepts such as 'the self', 'memory', 'public action' and 'national identity'. But before that it is perhaps useful to start the discussion by presenting the connections people made between the three concepts.

Between comfort, belonging and commitment

'Commitment to home is commitment to the family, it is a sense of history, a sense of belonging, familiarity, emotional investment.' (Janet, 40s, single, British – Jewish, London, 19 August 1999)

'The more sense of belonging, the more comfortable and the more committed I am.' (Mandy, 28, single, Canadian – Indian, Jerusalem, 16 June 2000)

'Commitment means keeping the house so that it provides comfort in the sense of keeping it tidy, having food [which also gives a sense of belonging] . . . to keep it in the best way. This is my commitment . . . to take care of these things.' (Magda, 30s, single, Palestinian – Israeli citizen – minority, Jerusalem, 25 March 2001)

For Janet, Mandy and Madga the meaning of commitment is connected to comfort and belonging. For Janet commitment is associated with family ties and belonging. For Mandy and Magda commitment is associated both with belonging and comfort. It seems that for many people commitment is the act, comfort and belonging are the essence. Being committed to one's own comfort and sometimes one's own sense of belonging one acts to fulfil these needs and desires (Mesch and Manor 1998). 'Commitment to what' is then constructed in one's own identity and one's own reflections on the meanings of comfort and belonging.

This understanding of the connection between the three concepts can be associated with people's involvement in public actions, which are related to planning and development. Commitment then becomes the driver and the motivation of people's activism to fight for or against development projects that fulfil or work against meeting their sense of comfort and perhaps even their sense of belonging to the different categories of their environment. Let us start to deconstruct this notion as it has been identified in people's narratives by looking first at the links between people's commitment to themselves and their sense of commitment to the environment.

Commitment to the self – commitment to the environment

'*I don't feel commitment to* my home, *it's my mom's owned home.* The commitment is to myself *and not to the home, the home provides [aspects of] commitment to myself.*' (Linda, 30s, single, Israeli – 'nomad', London, 30 July 1999)

'*The only place [home] that I can really* be myself *and do what I like.*' (Robert, 30s, single, British – white, London, 1 September 1999)

'The street *has to be* committed to me *in the sense that it will provide me with space. The street has to be beautiful because I pay taxes and it needs to be expressed in my environment.*' (Aziza, 30s, single, Palestinian – Palestinian (citizen of Israel), Jerusalem, 7 August 2000)

'*I have a problem with Jerusalem that it is changing,* the city is not committed to me, *secular parts are becoming religious and create hostile environments . . . all the young people leave because of their disappointment with the city. It is not a place you can trust, the city is not faithful so it is hard to be committed to it.*' (Rebecca, 30s, married, Israeli – Ashkenazi, Jerusalem, 3 February 2000)

Linda, Robert, Aziza and Rebecca, citizens of London and Jerusalem, distinguish commitment as a 'give and take' process between the self and the environment. 'The home provides commitment', 'the street has to be committed to me', 'the city is not committed to me'. The home, the street and the city become almost human entities that the self interacts with. Commitment becomes a component in the relationship between the self and the environment, a 'contract' that each side has to fulfil. The more 'public' the environment the more difficult it is to fulfil this contract. At the home level, it is a contract between oneself and one's home or family members and their home. Commitment to the home derives from one's own basic commitment to oneself, to one's own needs and desires. The home is an environment that allows one to be oneself, to grow, to relax, and thus one is committed to that space. This is usually a physical commitment: keeping it tidy, cleaning it, making it nice and comfortable, or providing the basic conditions so that the home remains an environment where one can be oneself and feel a sense of comfort and of belonging. These are basic emotional needs and thus the relevance and connection between comfort, belonging and commitment at the home level are much clearer to people and easy to associate with.

In public spaces this 'contract' between the self and environment can be labelled city citizenship or city residency. People pay taxes and expect 'the other side' – the authorities, to fulfil their obligations to themselves. This is why people say that the street and the city have to provide an adequate sense of comfort and belonging as part of this citizen contract and in order for people to feel a sense of commitment to the street or the city. This 'contract' can include an adequate level of services and city management and development so that citizens' needs for comfort and belonging are maintained. The management of the street and the city is perceived as the duty of 'the authorities' to the city citizens. When people feel disappointed in the politics of city management, as reflected in Rebecca's narrative, they find it hard to be committed. This sense of disappointment is because the authorities have not fulfilled the terms of 'the contract'. These terms could be concerned with notions of equality, trust, democracy or any other norms of citizenship. Commitment then becomes part of citizen–authority relationships and 'the right to commitment' is the right to demand the fulfilment of the citizenship contract.

Commitment to a place is associated with making choices. Where you choose to purchase your flat or home is where you choose to commit yourself to that 'citizen contract', it is where you choose to become a part of the collective that live there. hooks (2001: 438) chose to live in New York, she made a commitment to the city by making it her home: 'By purchasing a flat', she writes, 'I feel that I am making a commitment to life in New York City'. As she notes, this move was for her a change in her attitude to the city, which until then was 'a place where one gets lost' and 'Home as I understood it was a place where one would never be lost' (p. 438). This act of purchasing a flat, a home in the city, was for her an act of commitment to life in the big city in spite of her resentment. It may be a latent commitment with no public expression but the moment she purchased her flat she signed this contract, she pays taxes, she can vote, she is affected by the city planning and management in her everyday life.

Commitment to the city as a choice-making procedure has its spatial empha-sis in the process of out-migration from a city, which sometimes reflects a lack of commitment in addition to other reasons, such as a search for employment. Indeed much research that deals with people's motivation to emigrate focuses mostly on economic motivation, which is probably the most significant. However, it is argued here that even behind these explicit economic reasons there could sometimes be hidden sentiments of disappointment, of a lack of commitment to the city, especially in Jerusalem.

This perhaps explains the negative migration balance in Jerusalem in the last decade, which increases every year. In 1991–96 the negative balance was –6,000 on average per year. In 1997 it increased to –7,600 people and in 1999 it stood at –8,000. This negative migration balance is higher than in any other city in Israel (Jerusalem Institute for Israeli Studies 2001). This means that 12.5 people out of 1,000 left the city (Central Bureau of Statistics 2000). People leave Jerusalem for a variety of reasons. Job hunting is probably one of them but it seems that out-migration is also to do with resentment regarding city politics of planning and development, which explicitly discriminate against Palestinians and favour ultra-orthodox neighbourhoods. Like hooks, who associates purchasing a flat with an act of commitment to the city, it is hard for people to make a home in a city to which they do not feel a sense of belonging. The reasons for people's lack of commitment are reflected in some of the nar-ratives of people living in Jerusalem. They have not yet left the city but their disappointment becomes more and more explicit.

Staying or leaving the city is associated with commitment in Mandy's narrative. For her staying or leaving is associated with staying in relationships that are complicated. Here again commitment to the environment is associated with commitment to the self or selves:

> 'Commitment to the city means creating a sense of staying, to make it a better place to live – the same as in relationships.' (28, single, Canadian – Indian–Muslim, Jerusalem, 16 June 2000)

Commitment is perceived as an essential part of human relationships or as mutual responsibility. Relationships between human beings become a metaphor for relationships between individuals and the environment. 'A sense of staying' is probably a reflection of a sense of commitment to the city. In Jerusalem this sense of staying is much more contentious than in London.

In London, commitment as making choices and the reflections concern-ing disappointment with the city's management were not as explicit as in Jerusalem. It seems that most people interviewed did feel this 'sense of staying' in London, which is missing in Jerusalem. Lynne who is Scottish and who has lived in London for the last 20 years was the only one to express reflections and doubts about life in London. She said she was waiting for the Mayor's election (which took place in 2000) and its effect on life in the city before deciding whether to stay or leave London. It is not that everybody in London is totally happy with life in the city. People in London did criticise aspects of the life and environments such as the city centre and public transport, but it seems that for Londoners what annoys them does not cause a lack of commitment to the city as happens in Jerusalem. These different reactions in the two cities emphasise the multi-layered and rich expressions of this notion in different urban contexts.

Commitment as public action

'I wouldn't like my street to become noisy, congested or have too many buildings so I would stop planning permission to build a factory in my street, I would fight against it.' (Robert, 30s, single, British – white, London, 1 September 1999)

'If I see something that I don't like I try to change it, I am worried about the environment. That "they" will build a polluting industry near my street or in the wadi *nearby that we consider an open space.'* (Michal, 40s, married with five children, Israeli – Anglo-Saxon–Westerner, Jerusalem, 27 September 1999)

For people in both London and Jerusalem public action and involvement are associated with their sense of commitment to the place they live in. People would fight against any changes in their near environment that would have negative effects on their lives. The 'polluting factory' came up in Robert's narrative but was repeated quite often as an example of a change in the environment that most people would fight against. It seems that the 'polluting factory' is the environmental nightmare of most people in both cities, especially in their neighbourhood, and it symbolises the 'red line' of environmental problems that would motivate people to become active and object to municipalities' or developers' intentions.

In London, people mentioned commitment in respect of their wishes for the neighbourhood to be improved. Stuart mentions his commitment to the neighbourhood's improvement by taking part in the planning committee:

'Commitment to my neighbourhood is that I want it to be improved. The library, the cinema. If I could do it I would do it because I am committed. I could go to the planning committee but I don't think I am qualified.' (Stuart, 30s, British – Ghanaian–Afro-Caribbean, London, 6 August 1999)

Other people in London also mentioned fighting against demolishing old buildings as an act of commitment. Janet says:

'I would fight to keep buildings and against the MacDonald's invasion.' (Janet, 40s, British – Jewish, English, London, 19 August 1999)

People in Jerusalem mentioned this association between commitment and old building preservation or green areas preservation as well. Some people in Jerusalem mentioned the action taken to protest against the municipality's intention of building houses in part of the Jerusalem Forest as an example of their commitment to preserve green areas. This is a beautiful forest located in the Hertzel Mount that serves as a place of recreation, picnics and outings for many Jerusalemites. Lately several developers have initiated plans to construct housing units in the forest area with the support of the municipality. Liat and Neta associate commitment to the city with their action against such projects:

'I would object to city plans. I am committed to city life . . . to quality of life in the city: the buildings, green areas, the maintenance of the city.' (Liat, 40s, single, Israeli – Jewish, Jerusalem, 31 July 2000)

'I feel committed emotionally [to the city], to read and to know what's happening, to appeal to plans to destroy the forest of Jerusalem . . . not to let it be destroyed.' (Neta, 50s, married with three children, Israeli – Ashkenazi, Jerusalem, 11 June 2000)

Perhaps the most explicit expressions of commitment to the neighbourhood as public action are in Ronit's story in Jerusalem and the Bangladeshi community in London. Ronit is one of the active members in the *Kattamonim* – her low-income Jewish neighbourhood. Their activism is directed against another developer's plan to build skyscrapers in an area that now functions as a natural park. The area was originally leased to one of the kibbutzim near Jerusalem for developing a citrus industry. Because of the changing priorities from agriculture to other fields of economic activity the kibbutz ceased to grow citrus in the last decade or so. Natural vegetation slowly grew in the area and it even became a habitat for some deer herds. As a result the area has become a recreational and picnic place for the residents of the low-income, disadvantaged neighbourhood near where Ronit lives. No wonder that the municipality's intention of supporting the developer's plans to build skyscrapers and demolish parts of this natural park caused many objections. Ronit tells the story of this public protest with connection to the meaning of commitment to the neighbourhood:

'I am committed to this neighbourhood, my family lives here, . . . it was a tough neighbourhood but it gets better . . . the park nearby creates a sense of belonging to the neighbourhood . . . there is nature there, animals, it is our place of recreation, a place to recharge batteries, all the people in Jerusalem come to this park . . . in our neighbourhood there are no green areas so we go there and it's beautiful . . . so I'll do everything to keep it green and prevent building 1150 flats and roads in this area . . . I'd go to demonstrations. We have already organised two demonstrations, we signed petitions, and we are trying to convince the regional planning committee to leave it green; we organise activities with other organisations as well.' (Ronit, 40s, divorced with three children, Mizrachi – Mizrachi, Jerusalem, 8 June 2001)

The developer's intention of building in the natural park area are what makes Ronit 'go out to the street'. This is for her the association of commitment to the neighbourhood. This kind of active commitment is not evident in every person. Ronit herself mentions the difficulties of making people become committed and active:

'There is a basic discrimination here [because it's a disadvantaged area], if "they" [the developers] dared to do the same project in a rich neighbourhood the people would fight even harder. The problem is not only with the developers, the problem is with the people in the neighbourhood, they are a bit indifferent because they don't earn a lot of money and life is tough for them and they are old.'

Ronit's narrative highlights the problematic position of disadvantaged neighbourhoods in their fights with the municipality but also the difficulties in realising a sense of public commitment among the residents of these neighbourhoods, because these people have to fulfil their family commitment first. Is this connection between standard of living and a level of public commitment indeed relevant?

The Bangladeshi community is also one of the poorest communities in London and still their reactions to the developers' intentions in the Bishopsgate project in the East End seem to be explicit and practical. The Bangladeshi have organised themselves into community groups with the help of white leftists to object to the developers' plans in the area, which mainly targeted the construction of more office sites there. They established an umbrella organisation called the: Spitalfields Community Development Group to work with the developers to produce a Community Plan. Here there was a strategy among the developers to initiate a dialogue with the community. In return for community agreement to the project, the community representatives put requests to the developers regarding jobs' training and proper housing (see Chapter 4). The Bengali argued that they were willing to cooperate with the developers because they could not stop the development so they thought that they had better benefit from it.

These two examples of commitment as public action reflect the differences that exist between civil society's mechanisms in the two cities. In Jerusalem the fight to save the park is carried out with the involvement of various NGOs and neighbourhoods activists but the developers do not involve the communities in making development plans as much as they do in London. There, communities' objections to developers' intentions seem to be taken more seriously than in Jerusalem, where such procedures as making community development plan are not so familiar.

Commitment is associated with political action as well. In Jerusalem it is mainly against the discrimination against the Palestinian residents of the city. Rachel feels committed to fight against what she sees as an unjust situation:

'There is a quiet transfer [of Palestinians] . . . I am committed to political things . . . it is a charming city and I very much like living here and not in Tel Aviv but in spite of the city being aggressive and alien – I love it!' (Rachel, 30s, single, Israeli – European, Jerusalem, 24 March 2000)

For Rachel commitment to the city comes from her ideology. She sees commitment as fighting against what she terms aggression and violations against the Palestinians. Other people mentioned commitment with associations to 'justice', 'human rights issues', fighting against violation, etc. Here it is commitment to values, norms and ideologies that should be implemented in certain spaces, especially the city centre – a mixed space where people of diversity interact. In such environments norms of control, exclusion and power relations are expressed very explicitly and thus a sense of justice is becoming a commitment to the city.

In London such an act of commitment is more to do with the politics of planning and development. The white leftist in the East End fights together with local communities against the developers' intention to destroy Spitalfields Market and other sites in the area. Here the 'political' activism is more against the capitalist orientations of the city's development while in Jerusalem the political commitment is oriented against both the nationalistic and capitalist tendencies of the municipality in its development and planning.

To conclude this section, commitment as public action is expressed in both London and Jerusalem in fights against developers' intentions to change residential areas. In both cities these intentions harm mostly low-income neighbourhoods and make their residents fight against or negotiate over these intentions.

Commitment, memory and belonging

'I am emotionally and intellectually engaged when there are changes in architecture. London's city centre changes rapidly. I notice it and want the city centre to be the best place for people, more open spaces and more residential areas so that it is not empty at night.' (Stuart, 30s, single, British – African, London, 6 August 1999)

'It is important to me that the city centre remains authentic, that it will not be destroyed. I am willing to fight for it . . . I hate skyscrapers in the city centre . . . I want it to be a vacuum, to remain as it used to be . . . that it remains static like it used to be 30 years ago . . . that it will maintain its unique character even if the price is high . . . it is important to me that the city remains Arab in the Arab side of the city, that their development will be separated from our development . . . the way they want it and the way it suits them.' (Jacob, 30s, married with one child, Jewish – Israeli–Jerusalemite, Jerusalem, 17 June 2001)

Commitment to the city is associated for people in London and in Jerusalem with the city's history and its past. It is connected to memory, both personal and communal, that people wish to preserve. As already mentioned, commitment is the act of maintaining a sense of belonging of the Jewish and Palestinians in Jerusalem and of people of different identities in London. This is associated with old historical buildings preservation.

For others in Jerusalem commitment to the city is to its Jewish identity and belonging:

> 'A Jew like me is committed to Jerusalem. I am aware of the issue of sovereignty, of demography. I insist on living in this city, to take care of reasonable boundaries, that the decision making will not be only logical but also emotional, that is, logically perhaps it is better to give up the Old City but if "they" do it I will demonstrate against it. I distinguish between Israel and Jerusalem the same as a French man distinguishes between Paris and France. Not always does logical thinking lead actions and this of course creates tremendous problems . . . It must be understood that we will always live in conflict here, there will always be a conflict of interests . . . it is a complicated place but we shouldn't run away from complexities.' (Joseph, 40s, single, Jewish – Israeli, Jerusalem, 4 December 1999)

Commitment for Joseph is the act of living in the city as part of maintaining its Jewish majority. In Chapter 4 we elaborated on the fight for maintaining the balance of Jewish demography that the authorities initiate in Jerusalem. Joseph's narrative is an expression of this attitude on a personal scale. This is a commitment out of choosing, which is based on ideology. Joseph relates to the same 'sense of staying' that Mandy mentioned in the previous section. Both talk about commitment to a city, which will attract its citizens, but from different political points of view: Mandy wishes to change the current situation of discrimination and injustice and Joseph sees the current situation as a 'must' to keep Jewish sovereignty in the city.

In London most narratives on commitment express a deep desire to belong by being connected to historical buildings, to preservations of 'the city spirit'. Commitment is an act to maintain 'the city of spirit': 'It is time to reintroduce into our thinking about cities and their regions the importance of the sacred, of spirit' (Sandercock 1998a: 213). Most people in London expressed their deep commitment to the city as Stuart phrased it in his narrative at the beginning of the section. James also expresses his passion and commitment to the city. He says:

> 'I am totally committed to the city. I have a fierce interest and a deep political involvement in the politics of the city.' (James, 60s, British – Western–English, London, 1 September 1999)

For James commitment to the city is associated more with current political involvement than with past memories. James seems as passionate about London as Joseph is passionate about Jerusalem in his narrative. Their expressions of their passion and commitment to the city are based on different reasons, but they are still committed.

How can cities fulfil our passions and desires and perhaps by that make us more committed? Sandercock argues that this could happen if cities have enough public spaces and places that people can appropriate or 'privatise' for their own individual use. These are the 'sacred places' in urban landscapes that probably differ for different people and different identities: 'places loaded with visual stimulation, as well as places of quiet contemplation, where we can listen to the "noise of stars" or the wind or water, and the voice(s) within ourselves' (Sandercock, 1998a: 214). It seems that looking at the notion of memory as a component of urban planning is an important step in creating such sacred places in our cities.

Commitment and national identity

'What is commitment to your country?' I asked the people I interviewed. This question emphasises an interesting and important scale of 'the country' which can be compared to notions of commitment to other categories of space. Commitment to one's own country evolves issues of belonging, citizenship and nationality and clarifies the connections between them. People's responses to this question probably reveal their sense of inclusion or exclusion in their own society and country.

Commitment to one's own country is strongly connected to one's own national identity. In Israel, commitment to one's own country seems the most explicit and clear-cut issue to identify with among the Jewish people:

'I have high commitment to the country even though it's a commitment deriving out of criticism. My criticism derives from high commitment to the society in the country I have to be part of. This is the highest level of commitment in terms of my activities as compared to other levels of space . . . we perceive our national space as a very intimate space and this is one of the strong issues in our society . . . In London people are very active in their councils because that is the scale of influence and belonging there, and here I belong to the nation . . . this is the Home!' (Eitan, 30s, married, Israeli – Mizrachi, Jerusalem, 3 February 2000)

'Commitment to the country derives from one's own willingness to identify. For me it is impossible to identify with this country . . . we perceive the country as an extended neighbourhood in the sense of our care and interest in the country.' (Rebecca, 30s, married, Israeli – Ashkenazi, Jerusalem, 3 February 2000)

'Commitment to the country is thinking about the future and whether what I do now will improve the situation in the future.' (Emanuel, 50s, married with three children, Israeli – Jewish, Jerusalem, 23 February 2000)

The Jewish people in Jerusalem perceive commitment to the country as commitment to an extended home. It is interesting how 'the country' is perceived as intimate a space as 'the home', and as much as people care about their home they care about their country in a much more intimate way than they care about their street or neighbourhood. Commitment here is again about action, being involved, changing and improving existing situations. Another expression of commitment of the Jewish people is the fact that they stick to the place, to the country, in spite of their criticism of the ways the politics of the state and city management are handled. They may leave the city, as we have already seen, but they stick to the state.

Among the Palestinians living in Jerusalem commitment to the country is a very complicated issue, both for those who are officially citizens of Israel and for those who are residents only. Most Palestinians said that they do not feel committed because they do not feel that Israel is their own country, in spite of their having Israeli citizenship or residency. Perhaps the most striking reaction is Aziza's response. She is an Israeli citizen, but when I asked her 'What is commitment to your country' she looked at me and said: 'This is a very complicated question to ask', then sadly answered: 'I have no country!' Most of the Palestinians I interviewed felt the same:

'I don't feel that this is my country. I don't feel like an independent citizen that I can fulfil myself and my identity.' (Fatma, 40s, single, Arab – Muslim (citizen of Israel), Jerusalem, 18 April 2000)

'In our situation today the commitment is to keep Jerusalem as Palestinian as possible by staying in it – a matter of demographic commitment.' (Amal, 40s, married with two children, Palestinian – Arab–Palestinian, Jerusalem, 7 August 2000)

'There is a problem here. I don't feel that this is my country. I don't feel I belong to this county so how can I feel committed . . . my parents brought us up to feel commitment to the state, to feel that this is my country and that everything will be OK, but my personal experience is different and I feel like my parents lied to me and I am angry with them because they didn't tell me that the state is not committed to me [Arabs] as it is committed to the Jews. Because we are Arabs we get less but my parents didn't tell us this is not right. The state doesn't give to me, just demands from me, to say that I am faithful and committed and I am sick of it . . . like a never-ending story . . . I do not want to be part of this game.' (Magda, 30s, single, Palestinian – citizen of Israel, Jerusalem, 25 March 2001)

Fatma, Amal and Magda all express in their own words deep feelings of a lack of belonging and commitment but even more feelings of discrimination, as if they were deceived by the country that supposedly provides them with citizenship and equality. This is obviously not the 'full membership in a community' as Marshall defines the term citizenship (1950, 1975, 1981), it is rather the more complex and sophisticated interpretation of the notion, which creates nuances of different forms of exclusion and discrimination.

In London there are also different definitions of commitment to one's own country between 'the majority' and 'the minority'. Here are several narratives of the white British:

'England? No, I am committed to the UK, as my grandparents were Scottish. I am patriotic but at the same time I respect other cultures, so I am not a chauvinist – I am not sure what "British" is as it is such a diversified society.' (Robert, 30s, single, British – white, London, 1 September 1999)

'I don't think I could fight for the country. I am committed to voting, to have an opinion about matters and resources available. I'd miss the familiarity if I moved to another country.' (Richard, 30s, married, British – western, London, 1 September 1999)

'Commitment to my country is appreciating the good qualities of the country, being well understood.' (John, 60s, married with two children, British – white–Anglo-Saxon, London, 11 August 1999)

'I am committed to preserving the values of political democracy, human rights, my inheritance as part of Great Britain.' (James, 60s, single, British – Jewish–cosmopolitan, London, 22 August 1999)

It seems that the British white sense of commitment to their country is softer than that of the majority of Jewish living in Jerusalem, perhaps because basic issues concerning civil society are much more mature in London than in Jerusalem. John and James talk about commitment to values of democracy and 'the good qualities of the country'. They both reflect on commitment to the positive aspects of civil society in the United Kingdom.

The voices of the 'minority', that is, immigrants are different. Here for example is how Suna perceives commitment to her country. She is in her 40s and has lived in London for the last 15 years. She identifies her nationality – Egyptian, and her ethnicity – Mediterranean:

'Which country? I don't feel the UK is my country. Psychologically I feel I belong to my country of origin but I don't do anything political . . . I don't belong to any secular political movement which is pro-social justice, anti-corruption in my country of origin.' (London, 29 July 1999)

Commitment for her is also about action, fighting for change, mostly political change. She seems ambivalent about her understanding of commitment and her actual life in the United Kingdom, which do not reflect any commitment. Lynne is a different type of immigrant. She is Scottish, white, and has lived in London for 20 years. In spite of the similar cultural contexts and language between her home in Scotland and her life in London Lynne does not feel committed to Britain:

> *'Scotland is my country and then Europe, I would skip British identity. Funny position, hey. I do not abandon the idea of being Scottish.'* (Lynne, 40s, single, Scottish – white, London, 29 August 1999)

Stuart says that he lives in 'layers of commitment', an expression that describes the situations of most minorities or immigrants. He is not an immigrant, he was born in Britain, he is in his 30s and defines himself as British – Ghanaian–Afro-Caribbean. He says:

> *'I have layers of commitment. On one layer, I am committed to the country I live in which means I am concerned about social and political movements in the country. Another layer is cultural, that means I am committed to cultural movements, poetry, music, and these two layers are my sense of pride, which makes one committed. The other layer is the mythical legendary layer – my home country in Africa. It means to be committed to a country I don't know directly but it contains all mythical assumptions I grew up with. It is caring about what happens in your country.'* (London, 6 August 1999)

Although born in Britain, being black makes Robert feel connected and committed to two cultures: the Western culture and his African culture. Here issues of culture and perhaps race emerge. Robert feels committed to his roots, his imagined culture and homeland. Multi-layered commitment as Robert describes it can be associated with notions of 'multi-layered citizenship' or 'citizenship of the global city' that were identified in Chapter 3. Both terminologies emphasise the different layers and contexts in which the notion of citizenship is expressed. Issues of equality and difference seem to be part of citizen identity in the global city and what Robert identifies as 'layers of commitment' can be associated as part of citizen–state 'give and take' relationships. Citizenship in the global city is about receiving equal treatment but also about being able to maintain different kinds of commitments to other cultures.

Sentiments of commitment to one's own country are complicated both as related to Israel and the UK especially among immigrants and minority groups. Although issues of discrimination, which are associated with different nationalities, are more explicit in Jerusalem and perhaps Israel, people who live in London expressed ambivalent feelings regarding commitment to the UK and London if their national identities are different then the White – British. Sense of commitment to one's own country seems to be independent of the city one lives whether the global city or the holy city it is rather more dependent on one's own national identity.

Summary and conclusions

This chapter analyses the multi-layered nature of commitment to the environment. The analysis of people's narratives emphasises how commitment can be personal or public, passive or active, physically oriented or socially and ideologically oriented. Its multi-dimensional character made it more complicated to define out of the three concepts. If people could relate to the notion of commitment it was either to 'the home' or to 'the city'. Commitment was then usually associated with its physical aspects, that is, to keep a place (home, building, street, etc.) tidy, clean, not to make a noise, not to disturb other people. Here are some of the main issues that came out in people's narratives:

- *Commitment and the 'self'*: Commitment has a strong association with 'give and take' relations. Commitment to the 'self' is what affects the level and intensity of commitment to the different categories of environment. The extent to which people feel that the home, the street or the city is committed to them influences the level of their commitment to these environments. This can be seen as a 'private' or individual expression of the 'citizenship contract' between the citizens and the authorities. The street and the city's commitment to the 'self' are in fact the authorities' commitment in such environments. Once the authorities' commitment to maintenance and development of such public spaces are fulfilled people feel committed. (This can include maintaining public spaces, keeping them tidy and clean, taking care of any physical obstacles, etc.)

- *Commitment and making choices* have their expressions in people's explicit or implicit decisions to live in a city. An explicit expression of such a decision can be purchasing a home in a city. Lack of commitment can be expressed in out-migration. In Jerusalem more people associate commitment to the city with such notions of staying or leaving the city than they do in London. This 'sense of staying' in the city is a sentiment, which is developed out of this 'give and take' process between city citizens and the authorities.

- *Commitment as public action* is one of the dominant associations of people in both Jerusalem and London. In both cities people perceive their public actions as engaged with objections and protests against developers' intentions to reshape their environments. In both cities the developers' intentions have negative effects on low-income neighbourhoods, where the residents organised themselves to object to these projects.

- *Commitment to collective memory* is usually related to the preservation of old historical buildings. This can be associated with the desire to maintain 'the city of spirit' that is partly based on collective historical memory and is spatially expressed in the sites where these histories took place.

- *Commitment and national identity* can become a sentiment with very strong positive associations, such as among the Jewish people in Jerusalem and the white English in London. Such a commitment can also mean becoming active against nationalistic or capitalistic intentions of the authorities. The actual staying in the city and in the country in spite of their criticism is perceived by many as an act of commitment. For the Palestinians in Jerusalem commitment to one's own country is a complicated position and causes conflict, both internal conflict, between one's commitment to oneself and one's commitment to other values or people, or external conflicts between the Jewish and Palestinians. In London it seems that immigrants, even those who have lived in the city for many years, do not feel as committed to the country as the white English. 'Layers of commitment' is perhaps the expression that explains the situation of minority and immigrant people regarding this notion of commitment to one's own country.

- *Commitment is the act – comfort and belonging are the essence.* These are the relationships between the three elements, as the people perceive them. Commitment is perceived as the act of reaching the desired levels of comfort and belonging in one's own environment.

'The right to commitment' becomes an expression of the citizen–authority mutual responsibility or their 'give and take' relations. This means that 'the right to commitment' is the right of people in the city to receive what they ought to from the authorities and to give what they ought to give for the maintenance of a healthy and just society. The more elements that reflect people's identities are built up in urban spaces the more committed they feel. For some these are cultural elements such as museums; for others they are old historical buildings, green areas, etc. These elements also include norms and values of equality and democracy, which are connected to 'the politics of planning and development'. As such the right to commitment becomes a part of citizenship rights and what is termed in Chapter 3: 'spaces of citizenship'. Citizenship is about a mutual commitment to create spaces of citizenship in which principles of equality become spatial. The situation of the Palestinians in Jerusalem can be perceived as a description of 'spaces of discrimination', where the basic terms in the citizenship contract are not fulfilled. The Palestinians pay taxes but do not get the level of services that the Jewish people get. 'The right to commitment' in public spaces is the right to fight for such spaces of citizenship to function. It is the right of every city citizen to demand the services and infrastructure that they are supposed to receive as part of their citizen contract.

Note

1 The use of 'environment' here refers to the six categories of space identified in Chapter 1: the home, the building, the street, the neighbourhood, the city and the city centre and urban parks.

Gender identity and the local embodied knowledge of comfort, belonging and commitment

'What is the meaning of the city? It is civilisation, change, colours, and movement . . . I am originally from a village so that helps me to appreciate the city . . . but in East Jerusalem the space is not mine, it is like going back to my village. In a period of deurbanisation that East Jerusalem is undergoing . . . There is a tendency to return to traditional social structures of the village. Deurbanisation means an acceleration of the patriarchy, the rejection of the 'Other', the different . . . and then family norms are strengthened.' (Aziza, 30s, single, Palestinian (citizen of Israel), Jerusalem, 7 August 2000)

Does gender identity make any difference in people's everyday life experiences? Do men and women define comfort, belonging and commitment differently? This chapter aims to deal with these questions. After a detailed analysis that focused on each of the themes separately (Chapters 5, 6 and 7) this chapter aims to examine whether identity issues, that of gender, matters in people's daily experiences regarding the three related themes. The decision to write a chapter focusing on gender issues rather then examining gender identity in the chapters derives from the idea that it is important to do an analysis that works across the themes in addition to the analyses presented in previous chapters, which focused on each of the themes separately.

Back to Aziza's narrative. Her embodied everyday life experience emphasise the connection between the situation of women and 'the politics of city planning and development'. Her experience shows that the more degraded and the less developed a city is, that is, the more abuse there is of civil rights, the more harsh and exclusive a place it becomes for women and for the 'Other'. Aziza's narrative emphasizes how identity issues embodied in her knowledge, in this case gender and nationality, cannot be disconnected from issues of urban planning and development.

Three dominant themes have been identified in analysing men's and women's narratives from gender perspectives: first, the ways in which the division of gendered roles within the household reflects experiences of comfort, belonging and commitment; second, the expressions of gendered power relations in women's narratives; third, the gendered symbolic construction of spaces – 'the forbidden' and 'permitted' spaces and the role of clothing in women's free movement in urban spaces.

The connection between households' gendered division of roles and experiences of comfort, belonging and commitment

Domestic and public spaces

> 'The choice of house makes me feel I belong and also the ritual of cleaning it when I first moved in, which makes me feel that it belongs to me. Without it I wouldn't feel comfortable at home.' (Linda, 30s, single, Israeli – 'nomad'– Western, London, 30 July 1999)

> 'Cleanliness and order . . . make me feel comfortable at home.' (Aliza, 30s, single, Israeli, Jerusalem, 23 April 2000)

Experiences of comfort and commitment *at the home level* are much more affected by the gendered division of roles than notions of belonging. Linda's narrative connects cleaning of the home to belonging but also to a feeling of comfort at home. Most women expressed their sense of comfort and commitment to the home as connected to their reproductive role and being in charge of maintenance, cleanliness and order – even those who are single. Moreover, when talking to two couples in both cases women's embodied knowledge of *commitment to the home* was associated with cleaning, house maintenance and order, whereas for men it was much more difficult to define commitment to home, or their definition was connected to issues other then reproductive roles and maintenance. It seems that the notion of 'order' is very much connected in peoples mind to that of comfort. For women it is more related to the domestic–home base. Men also relate to order at home, not as their duty or responsibility but more as one element that makes them feel comfortable and even belonging.

Men related to order as providing comfort in public spaces such as streets or the city. Spatial order raises one's familiarity with the city. Richard emphasises order as knowledge in his narrative on comfort in the city:

> 'Comfort in the city is that I know it. I hate feeling lost. I like being surrounded by people.' (Richard, 30s, married, British – Western–English, London, 1 September 1999)

Abraham mentions order, which is more associated with planning perhaps because of his professional identity as an architect who deals with the 'arrangement of space':

> *'Comfort in the street is first and foremost order, a basic order . . . that cars will park where they should park and pedestrians will walk on pavements . . . a clear division of order between different uses of the street.'* (Abraham, 50s, married with three children, 29 February 2001)

Familiarity, order and knowledge of the streets was repeated in men's narratives in both Jerusalem and London more than in women's narratives in either city. It probably does not mean that these notions are less important for women but simply that they were not mentioned in their narratives.

The role of motherhood

As a result of their gendered roles women, especially mothers, use the environment near their home more intensively, especially for shopping, taking the children to school or walking with their push-chairs. Their roles as mothers are one of the significant aspects of their embodied knowledge as related to all the three concepts. Here are some narratives that demonstrate these connections:

> *'I began to feel* comfortable *in the neighbourhood only when my daughter was born, then I started to walk in the neighbourhood, to get to know other mothers who live close by. We usually meet and go together to the nearby park.'* (Amaliya, 30s, married with one child, Israeli, London, 22 August 1999)

> 'Comfort *in the neighbourhood is a matter of the stage in the life cycle, when my kids were young it was important what services the neighbourhood had – kindergarten, school, nursery, clinic, grocery, because we didn't have a car so everything had to be close by . . . today my kids are grown up I look for something else in a neighbourhood.'* (Ruth, 40s, married with two children, Israeli – Ashkenazi, 4 November 1999)

> *'The school of my kids makes me feel I* belong *to the neighbourhood, because everybody sends their children to school and kindergarten so the school belongs to the neighbourhood, but also the neighbourhood belongs to the school because we all belong to the same network. Parents bring flowers and cakes to the school, my husband painted the fence . . . The people I know in this neighbourhood are those I know from my kids' school.'* (Efrat, 40s, married with two children, Israeli – Tzfonit (from north Tel Aviv), 9 August 2000)

Parenthood and especially motherhood means more extensive use of one's own environment, especially those services that are directly connected to children's upbringing such as kindergartens or schools. The school was mentioned in many cases as the core of social activities in the neighbourhood, the spatial element that dictates its boundaries and activities. Thus more women than men associated comfort with having services or other central activities in proximity to their own residence, or even living in a mixed neighbourhoods of services and residences. Women more than men expressed these aspects explicitly precisely because it is such a significant part of their daily embodied knowledge constructed by their gendered roles.

Gendered power relations reflected in women's experiences regarding comfort, belonging and commitment

Power relations are expressed in men's and women's narratives in many ways, but especially when using key words such as 'control', 'freedom', 'safety', 'security' and 'privacy'.

A sense of control and freedom are two interrelated words that were mostly connected to the embodied knowledge of the three themes, but perhaps a more clear association has been made between comfort *at the home level* and different expressions of power relations. This is obvious if we return to the literature on power relations at home (Chapter 5), especially the work of Massey (1994), hooks (1991) and Sibley (1995) on this issue. Power relations and control are major elements affecting women's sense of comfort at home. As already analysed in Chapter 5, control at home means for most people control of which items or possessions should be at home but also about the order of those possessions. Some of the stories of Amaliya and Lynne living in London and Mary, Saida and Fatma living in Jerusalem appear in previous chapters, but it is worth presenting them again with emphasis on key words that connect the notion of power relations and women's sense of comfort at home:

> '*I feel much* discomfort *and* I don't belong *to the home because I live with my partner and* he has his own needs *and* his own tastes *which are different from mine. The way the house is arranged is not exactly how I would have arranged it.* It is too arranged, I don't like the furniture . . . *it makes me* feel less as if I belong. *Belonging for me is that* it is my own space *and that* I decide *what will be in it.* A total control.' (Amaliya, 30s, married with one child, Israeli (living in London), London, 22 August 1999)

> '*Comfort at home for me is about* freedom, *that everything is in* my control, *that there is nobody who can* suddenly disturb me *and* tell me *what to do and where to be. This is in contrast to my childhood when home was* a difficult place for me *with* a lot of control by my Dad.' (Mary, 40ss, single, Israeli – Ashkenazi–Jewish, Jerusalem, 4 August 1999)

'*Comfort at home is that* nobody can come in, that it's safe, it's mine, that I'm not sharing it with strangers, nobody watching me.' (Lynne, 40s, single, Scottish – white–Western, London, 29 August 1999)

'*Comfort at home means* my own privacy, I decide *what to do*, and I am free. *Discomfort means that* something disturbs this privacy, *unexpected guests . . . also having* other people living in the same house, *each has a different opinion so you are* not as free as you should be, *the house has* too many people *and then I don't have* my own privacy . . . *lack of it causes discomfort.*' (Saida, 30, single, Palestinian – Muslim–Arab, Jerusalem, 30 December 2000)

'I can not do what I plan to do. *For example, I want to work on my students' dissertations or to read some material or to write and* I can't do most of it *because I live in a house where* my mom lives *and she is the mom of all* my brothers *who come to visit her and there is such* a noise *that I just* can't work. *Also the kids want me as their aunt, they* want my attention . . . *It is* hard to draw the boundaries *in the village.*' (Fatma, 41, single, Arab – Muslim, Oriental–Oriental, Jerusalem 18 April 2000)

Five women, two Israeli–Jewish, one lives in Jerusalem, the other lives in London, another is Scottish living in London and two Palestinians living in Jerusalem. One is married, four are single. Different cultures, different stages in the life cycle but similar embodied experiences of power relations at home. Patriarchy has its representatives in their stories, the father, the husband and the family members ('the mother figure', the brothers), all representing valued cultures – western or non-Western – that put women in a weaker power relation position, which affects their sense of comfort and belonging. Patriarchy is an inherent component in many women's lives, either as girls or as adults. As already mentioned patriarchy raises its head when the city's planning and management and its network decline. Aziza identified this process at the beginning of the chapter. She experiences a much more rigid and fierce patriarchal atmosphere in the streets of East Jerusalem than before. It is not that the atmosphere in East Jerusalem before this last Intifada was totally inviting for women. Patriarchy existed then too but lately because of the city's decline Aziza feels that patriarchy is becoming more explicit and the life of Palestinian women more rigid and tough. Patriarchy is also visible in the streets of Banglatown in the East End of London in Muslim women's traditional clothing and head covering. In Chapter 3 I elaborated on this connection between patriarchal power relations and women's clothing and we will return to this point in the last section of this chapter.

A sense of control is also one of the major components that men mentioned in creating sense of comfort in their home. Men expressed their need to be in control of time, of possessions, of the order of things, but they never mentioned a situation of a lack of control, or a lack of freedom in their gendered embodied experiences and knowledge. These situations are apparently exclusively female! Perhaps the most striking and maybe shocking expression of the lack of freedom and control that Fatma feels in her home is when she says that comfort at home for her is: 'When I am tired and need my privacy I want my bed – this is home for me.' It is only when she is in her 'bed space' that she is not intruded on by her family members: mother, brother, nieces and nephews. This is another expression of her feelings of her home as prison (see Chapter 5). Here cultural identities interact with gender identities. Being Muslim un-married women both Fatma and Saida cannot live on their own and have to live with their families, against their will. Aziza has managed to escape from this difficult situation. She is single and expresses her feelings of comfort of living alone:

'*It is the first time I have lived on my own with no partner. It is a new experi-ence, it is very new for me to control space, my own space, totally, a lot of free-dom, fun, something that it is not known to me and I like it.*' (Aziza, 30s, single, Palestinian – Palestinian–Arab, Jerusalem, 7 August 2000)

The way Fatma reflects the dominance of control in her life by talking about what commitment means for her highlights the point made earlier on the dom-inance of power and control at home as related to the three themes:

'*Commitment means that "they" dictated for me what to do when I was young: to come back home before sunset . . . now it means family honour, not to have sex with men if you are not married, commitment is what they want from me . . . that the family knows everything about my life . . . to expose my privacy to them, not to live outside [the family home], to keep the village norms, to be modest, not to be exceptional as a woman, not to be a revolutionist, to respect old people, to be subordinate . . . but I do what I want . . . I am a revolutionist . . . privacy is privacy after all and I am trying to influence society to accept me as I am although I pay a high price.*' (Fatma, 40s, single, Palestinian – Israeli, Jerusalem, 18 April 2000)

Fatma's narrative reflects how commitment is perceived differently for men and women in different cultures. She describes in detail how restricted are what she perceives as 'the rules of commitment to her culture' that are dictated by her family and her society's norms. It is clear that for women in certain cultures such norms are much more strict and tough than they are for men.

It is at this point that the voices and the experiences of the Bangladeshi Muslim women would contribute to the analysis of this theme. Their missing voice is actually a reflection of the dominant patriarchal rules and norms that dominate their lives until now. When I asked my Bangladeshi male inter-viewees if I could interview their wives they said it was a bit complicated because their wives were not used to talking to strangers and also their English was not good enough. For the Bengali men to allow their wives to meet an 'outsider' was probably something that threatened the patriarchy in their soci-ety. The Bangladeshi men themselves raised another issue that emphasises the strong patriarchal control in Bangladeshi society, even towards young females who were born in Britain. It relates to the custom of arranged marriages of their children and especially their daughters. Some of them told me how hard it is becoming to stick to this traditional rule because the daughters are becoming more educated and more British and therefore the options of 'good' husbands back in Bangladesh are becoming less desirable for the daughters themselves. Now, the fathers have to find appropriate husbands among the Bangladeshi community members in Britain. This tradition of arranged marriage is another dimension of patriarchy, which is not spatially explicit but is identical to the relatively strong influence of tradition and patriarchy on Bangladeshi women and the low level of freedom and control they have in their lives.

The gendered construction of symbolic public spaces of comfort, belonging and commitment

Fear in public spaces: the construction of forbidden and permitted spaces

So far we have seen how comfort, belonging and commitment to the home are gendered constructions of power relations. But these constructions are not only relevant to the 'private' – the home space. As we see below, power relations and control also affect women's movements in public spaces, expressed in fear and a sense of safe and unsafe in the street, public transport and urban parks. Feelings of fear of using these spaces came out in both men's and women's narratives but they were more dominant in women's narratives in both cities.

Women's movements in space are very much controlled and restricted by cultural and ethnic norms and values, which in turn symbolise spaces as 'spaces of modesty and immodesty' that sometimes become 'forbidden' or 'permitted' spaces for women in specific cultural contexts. The 'cultural guards' of society, that is, men or elderly women, dictate the boundaries of these spaces (Fenster 1999a).

As discussed in Chapter 2, in 1999 I wrote about the cultural construction of space of Bedouin women living in the Negev, South of Israel (Fenster, 1999b, c). There I mentioned the construction of a public/private dichotomy as forbidden/permitted cultural constructs of space, which become restrictions for Bedouin women's movements in their towns. The narratives of women living in Jerusalem and London reveal that these terminologies are relevant not only for Bedouin women but for women in general.

The construction of spaces as forbidden and permitted is usually a result of fear. Fear in this context is gendered, more associated with women than with men living in the city. Fear of harassment in public spaces cuts across women's everyday experiences in London and Jerusalem. It cuts across other identities such as nationality, marital status, age, sexual preference, etc.

Feelings of fear and safety are sometimes very connected to the ways urban spaces are designed and planned. Here is Rebecca's story:

'The avenue in my street is scary because there is only one exit to it, you can't leave it from everywhere . . . And there are benches where weird "creatures" can sit and molest you and you feel trapped . . . so it is not so pleasant . . . If you get into the avenue you are lost . . . it is really a male planning, "they" did it because of the transportation but it prevents me from walking in the avenue.' (Rebecca, 30s, married, Israeli – Ashkenazi–Israeli, Jerusalem, 3 February 2000)

Certain spaces in the city can become a 'trap' for women, unpleasant and so unused. They become 'a planned trap', that is, planning created or designed those spaces so that they become 'traps' with no easy access, such as the avenue near Rebecca's home.

Urban parks have the same association. Some women perceive parks as 'hostile male areas': 'they are "stolen" areas. *I feel angry that I can't use them'* (Aziza, 30s, single, Palestinian, Jerusalem, 7 August 2000). What Aziza expresses here is mostly a sense of exclusion, but perhaps also a sense of a symbolic appropriation of space, of not being able to use certain spaces in the city because they are controlled and dominated by men. Eleonore talks about her sense of fear in urban parks. She says:

'Comfort in the park . . . depends on what time . . . At dusk I wouldn't go there. During the day I would go near where I live. Only there I feel safe. I wouldn't walk in the park in the evening.' (Eleonore, 50s, British – white–English, London, 1 September 1999)

It seems that fear makes most urban parks forbidden spaces after a certain time of day. Most women in both cities avoid using this space at night. Other research (Madge, 1997) shows the same attitudes of fear in urban parks, especially at night.

In London, women expressed the same level of fear and insecurity in urban spaces, especially public transport. Suna's narratives appear in Chapter 5, where we discuss 'comfort in public transport'. It is worth mentioning once again here with regard to the theme of forbidden and permitted spaces:

'Public transport is full of gangs of kids, people are drunk especially after football matches. When men are together they become aggressive and late at night when trains are empty it is scary.' (Suna, 40s, Egyptian – Mediterranean, London, 29 July 1999)

Suna feels that trains at night or after football matches are becoming 'forbidden spaces' for her in the sense that they are unsafe. Men in London also made comments about their sense of insecurity when travelling at night in empty trains. Men and women expressed the same feelings of discomfort and fear when they use the underground late at night.

What are then permitted spaces? Aziza's narrative on her embodied knowledge of what is for her comfort in her neighbourhood emphasises the nature of 'permitted space':

> '*I feel most comfortable in this neighbourhood because it is the most beautiful place in the city of Jerusalem. I am a person of constraints: I am a woman, Palestinian, alone [this neighbourhood is like] a microcosm – it reminds me of London: a variety of people . . . In such places I bloom, like a fish in the water, this is my sea. I feel very protected because this neighbourhood is on the border between West and East Jerusalem and it is the ideal place for me. I lived once in Rehavia [a Jewish neighbourhood] and felt suffocated. Here I can get easily to the Old City.*' (Aziza, 30s, single, Palestinian – Palestinian (citizen of Israel), Jerusalem, 7 August 2000)

A permitted space is an urban space, which allows Aziza to live as an anonymous person. This is a space of comfort for her. As a single Palestinian woman she acknowledges the constraints existing for women in her culture, constraints that are expressed in other Palestinian women's narratives. Aziza makes the connection between feelings of comfort in her neighbourhood and the feelings of freedom, perhaps cosmopolitanism, that she feels when she visits London. Here the neighbourhood in its narrow space becomes an imagined space of inclusion – a permitted space.

The role of clothing and the politics of planning and management in the city

Jerusalem – the city of 'holy' spaces

In Chapter 3 we discussed the role of clothing in expanding women's boundaries of 'permitted spaces'. In Jerusalem, clothing is one element that determines where women are forbidden to go who are dressed differently from the dominating cultural norms which dictate 'public behaviour' in these spaces. 'Symbolic imagined' spaces become realities in the sense that women's clothing plays a major role in making women feel comfort or discomfort, in particular within the ultra-orthodox neighbourhoods. All women living in Jerusalem, both Jewish and Palestinians, mentioned the Mea Shearim (Hebrew for one hundred gates) neighbourhood of the ultra-orthodox Jews as the place most forbidden for them (see Figure 4.7). In Jerusalem women's bodies and their clothing become central to their ways of moving in certain spaces. Here are the stories of secular women, Jewish and Palestinian, with special reference to the 'forbidden' space of Mea Shearim. All mentioned this place in connection with feelings of discomfort:

> '*It is very uncomfortable for me to go to Mea Shearim, it is hard for me to accept the authority of somebody who is extremist and rejects me from humanity. "They" will not accept me in my everyday clothing and I have to force myself to*

Figure 8.1 'Please Do Not Pass Through Our Neighbourhood In Immodest Clothes' – gated street in Mea Shearim, Jerusalem

adopt their identity and it is not comfortable for me. The same in churches or mosques.' (Sarit, 50s, married with three children, Israeli – Spharadic, Jerusalem, 22 April 2000)

'Mea Shearim is the least comfortable place for me. I can't dress the way I like . . . I like to walk there but.' (Susana, 30s, married with one child, Israeli – Jewish, Jerusalem, 13 July 2000)

'Mea Shearim is a place I avoid going. I don't dare because I feel that it is not only that I don't belong but it is like a gated place open only to the ultra-orthodox and only for Jews so I have never been there and didn't even think of going.' (Magda, 30s, single, Palestinian (citizen of Israel), 25 March 2001)

An ultra-orthodox neighbourhood such as Mea Shearim is seen as a gated community, a place associated by women as the most excluding place in Jerusalem. Its 'gates' are not physical but more symbolic. They have very visual expressions in the big signs that the residents hang at the two main entrances to the neighbourhood (see Figure 8.1). What we see in the photo is a visual expression of exclusion, targeted at women only. These regulations are highly visible, in some shops there is a clear sign saying: 'It is forbidden for women dressed immodestly to enter this shop' (see Figure 8.2). This is an example of a very clear and strict exclusionary boundary, which is directed to women only and is engaged with women's bodies as symbols of modesty. In these representational

Figure 8.2 'To the Female Customer: Please Enter My Shop with Modest Clothing Only'
   'Modest Clothing includes: long-sleeved shirt, long skirt – no trousers and no tight clothing', Mea Shearim, Jerusalem

spaces the power of patriarchy dictates women's free movement in space. It does it with very strict norms and rules that are mainly expressed in women's ways of dressing. These spaces are becoming 'forbidden spaces' for many women living in Jerusalem, both Jewish and Palestinian, unless they follow the rules of dress demanded of them.

What do ultra-orthodox women feel when they go outside their neighbourhood? The majority do not cross the neighbourhood boundaries because of issues of modesty. I managed to interview one woman who is not ultra-orthodox; she was secular and became religious. It is interesting to read her experiences as one who crossed identity boundaries. For her, watching people dressed 'immodestly' is perhaps as difficult as it is for secular women to be forced to dress 'modestly'. Hanna says:

> 'To walk in Jaffa Street [the city centre] . . . This is pritzut [obscenity] . . . Clothes that you are not allowed to look at. Everything is confusing . . . all bodies are exposed . . . the heart is crying for them and the fun [of going out to the city] is gone.' (Hanna, 30s, married, Jewish – Sepharadic–Israeli, Jerusalem, 26 June 2001)

Because of the issues of clothing and modesty the city centre is slowly becoming a 'forbidden' space by choice for Hanna, a place from which she consciously excludes herself because the norms and values of the streets' culture do not fit her own norms of body modesty.

Jerusalem's city planning and management allows such exclusionary norms to exist and function in spaces that are public with the support of the municipality. Mea Shearim is indeed a unique case in its extreme rules targeted at women's clothing. This is why many women avoid going through the main road of this neighbourhood in spite of its centrality in the city. These exclusionary rules make them into 'private' spaces in the sense that they are open to the community members only. Public spaces become 'private' in their seclusion character because of gated communities' norms, which gain the support of the municipality. Such spaces highlight the conflict between 'the right to the city', which focuses on equality, and the right of such communities to maintain their cultural and religious differences without offending 'the right of the city' of the rest of its citizens.

Similar cases[1] of such exclusionary constructed spaces are known in other places where ultra-orthodox Jewish people live. I have heard of two areas in New York state that became exclusionary to women because of the same norms. In one case, in Kiryat Joel, a town of some 15,000 ultra-orthodox Satmer believers located in the Northern part of New York state there is an area that is totally forbidden for women. This is where the Yeshiva (the school of religious studies) is located. If a woman wants to call one of her family members who studies there she has to ask another man to call him. The school is located at the top of a hill in the town whose population are only ultra-orthodox and that is why the norm is acceptable and maintained. The second case aroused more objections. In another part of New York state, in Square City, another group of ultra-orthodox people lives, this time in a neighbourhood in proximity to other non-Jewish residents. At the entrances to this neighbourhood there are signs similar to those in Mea Shearim in Jerusalem calling on women not to pass through their neighbourhood with immodest clothing. One woman who lived nearby appealed to the court, arguing that the signs abuse her civil rights. It is not known what has been the outcome, but what is perhaps important in this case is the fact that in different structures of civil society such as in New York state there are different understandings of the meaning of 'the right to the city'.

London – the city of diversity

In London there are many neighbourhoods populated by ethnic and cultural communities, and this brings to the neighbourhood services and shops that are significant to the culture of its inhabitants, so that some neighbourhoods become 'Muslim' or 'Indian' or 'Pakistani'. The residents of those neighbourhoods make sure that their religious needs are met by constructing mosques and temples. (see Figure 8.3). One of the famous examples in the last decade is the construction of the Hindu Temple Shri Swaminarayan Mandir in the suburban residential streets of Neasden. It is the first traditional temple to be

Figure 8.3 The Mosque in Whitechapel Road, East End, London

built outside the Indian subcontinent and it serves the growing demand of Hindu believers for such sacred spaces.

However, these ethnic and religious neighbourhoods are not transformed into 'private' exclusionary neighbourhoods such as Mea Shearim in Jerusalem. It seems that in spite of the religious differences and the strict rules that such religions maintain for women's clothing and public behaviour there is no single space in London that has such strict and clear exclusionary signs.

The Spitalfields–Brick Lane area in the East End of London is the area where the Bangladeshi established their community with all the facilities they needed to maintain a communal life. Walking in the Brick Lane area one notices the mixture of women's clothing. Some Bangladeshi women are dressed tradition- ally (Figure 8.4) but others are dressed in Western style. It is the same in other public spaces such as Hyde Park in London, which is used by a mixture of people such as the Arab Muslim population that lives nearby (see Figures 3.2 and 5.6) and people of other nationalities who also share these public spaces.

It seems that strict boundaries of exclusion in city neighbourhoods such as Mea Shearim are not possible within the framework of city planning and man- agement in London. As mentioned in Chapter 6, in many ways Mea Shearim has some similar characteristics to Banglatown, in that it also reflects a spatial sense of belonging which has a very strong spatial expression, as it does in Banglatown. Nevertheless, the sense of communal belonging in Mea Shearim is

Figure 8.4 Women in traditional dress, Brick Lane Road, London

exclusionary, especially towards women, the opposite of the welcoming spirit experienced in Banglatown.

We realise at this point how norms and values regarding intimate spaces such as the body are affected by the politics of planning and management of larger and much more public spaces such as neighbourhoods and areas in the city. The politics of planning and management, which advocate and support extreme segregation and exclusion, affect women first because their bodies are perceived as the expressions of modesty, honour and other religious norms. The more supportive the politics of planning and management are towards such exclusionary norms and rules the stricter and harsher is their effect on women's everyday practices.

Summary and conclusions

This chapter analysed the gendered daily experiences of comfort, belonging and commitment as they are reflected in men's and women's narratives in both London and Jerusalem. The narratives highlight the following issues:

* *Household gendered division of roles affect men's and women's sense of comfort, belonging and commitment* especially at the home level. Because

of their gendered household division of roles women associate their sense of comfort and commitment at home with their household duties. Men found it difficult to define a sense of commitment to home.

- *Order and spatial knowledge is much more associated with a male sense of comfort in public spaces then female.* Men's knowledge of the city's layout was the main denominator for their sense of comfort and belonging in public spaces. It was expressed more explicitly than in women's narratives.

- *Parenthood and especially motherhood increase the sense of comfort, belonging and commitment, especially to the neighbourhood.* Young mothers indicated that their sense of belonging and comfort to their neigh-bourhoods became stronger after having children. This is because of their intensive use of the space, especially in their daily walks with their babies. The use of education facilities as centres of activities in the community increases a sense of belonging for mothers, probably as a result of their gen-dered responsibilities within the household.

- *Patriarchal power relations are a major component in understanding notions of comfort and belonging at home on a gender basis.* The home is usually perceived in the literature as a safe place, a sanctuary for family members. Looking at women's narratives of comfort and belonging at the home level reveals that in many cases women lack comfort and belonging because of the patriarchal power relations they experience.

- *Patriarchal power relations are a major component in understanding notions of comfort and belonging in public spaces.* – Fear in public spaces is one of the major triggers in constructing symbolic, 'forbidden' and 'permitted' spaces in the city. Both men and women said they avoid using certain spaces in the city because of fear. Spaces such as streets, parks and public transport are sometimes becoming 'forbidden' for use by both men and women although they are more explicit in women's narratives.

- *The role of clothing and the politics of planning and management in the city* are very closely connected, as seen in women's experiences in Jerusalem and London. The more supportive the city management is towards notions of seclusion and segregation the more strict and patriarchal the norms of behaviour become, especially with regard to women's clothing in the city. The possibilities of using such strict rules in public spaces in Jerusalem can be seen as expressions of the complexities concerning the right to the city in Jerusalem, which are much more straightforward in London.

Note

1 The information regarding these cases was kindly provided by Professor Joseph Shilhav.

Different Ways of Knowing:
Diversity, Knowledge and Cognitive Temporal Maps

'*I visualise my childhood environment as a cycle. The street was a circle and I remember very strongly this circle. I remember a lot of travelling and a strong contact with nature; trips, journeys . . . I remember mountains, caves, trees, animals, sea, tents . . . these are the best memories of my childhood, the trips . . . a very nomadic lifestyle.*'

'*My present environment is very different from my childhood one. More static . . . like a big building which does not move . . . frozen.*'

'*My utopian environment? It is the opposite of the building I live in now, it includes more open spaces and greenery with the mountains as the boundaries, that is . . . no more houses, and behind the mountains there is the sea as if it's an Island.*' (23 April 2000)

These are three temporal verbal mental maps of Aliza, a Jewish woman aged 38 living in Jerusalem. Like all other interviewees she was asked to draw her childhood environment, her current environment and her future desired–utopian environment.

Aliza did not like the idea of drawing maps because she said she did not know how to draw so I asked her to describe the three temporal representations of her subjective environments. The differences between the three descriptions are quite amazing. The fresh and free description of her childhood, with emphasis on open spaces, nature, freedom, the 'outside' as opposed to the rigidity expressed in her present life and the utopian vision of the future that actually brings back childhood memories.

Raban writes (1998: 162): 'we map the city by private benchmarks which are meaningful only to us'. He calls it 'this private city-within-city' and inside this private city 'one builds a grid of reference points, each enshrining a personal attribution of meaning' (p. 163). It is this private, personal, symbolic meaning of spaces and places that are expressed in mental mapping, either verbal or pictorial. The way we map our private city is the expression of our symbolic construction of the city. We will draw objects that are meaningful to us and omit objects we do not use, care or perceive as important. Thus: 'the image is not the same for everybody, nor does it mean that same thing' (Pile 1996: 27).

Part III introduces these aspects of local embodied knowledge which are related to perception and images of space. It discusses these notions by analysing the images of comfort, belonging and commitment as they appear in what is termed here as Cognitive Temporal (CT) maps. These are cognitive maps of men and women's childhood environment, their present environment and their utopian environment.

CHAPTER 9

The images of comfort, belonging and commitment in cognitive temporal mapping

Mental perception of space, images and cognitive maps

Mental or cognitive perception of space is defined as a process composed of 'a series of physiological transformations by which an individual acquires, codes, stores, recalls, and decodes information about the relative locations and attributes of phenomena in his everyday spatial environment' (Downs and Stea 1973: 9). Mental processes relate to the inner capacity to literally map the world (Pile 1996). Lee (1968) suggests the schema – a psychological concept that helps in studying human behaviour in space. Head originally defined the schema in 1920 as 'a model of the pattern of nerve impulses that must be built up in the brain when we carry out any complex movement with our limbs' (Lee 1968: 97). Lee applies this concept to the definition of a neighbourhood and suggests that the schema of a neighbourhood is built up on a 'continuous input of sensory information from the physical and social objects in the urban locality, arising from our repeated transactions with neighbours, tradesmen, buildings, bicycles, parks, walls, etc.' (Lee 1968: 98). I will return to this point later in the chapter in the section dealing with belonging, but before that let us connect these definitions of mental, cognitive perceptions and schemata to the notion of cognitive map.

The notion 'cognitive map' is due to Tolman (1948), who demonstrated that rats, freely exploring an environment (maze), tend to construct an internal 'map' of it, which they later employ when needed to execute spatial tasks. In this paper, titled 'Cognitive maps of rats and men' he extended his theory to the human domain (Tolman 1948). Since then the concept has been employed extensively by cognitive scientists (psychologists, neurologists and the like) and also by students of culture and society, mainly as a reference to images of space or as a metaphor for space.

Cognitive mapping is what actually dictates our human spatial behaviour. It is based on cognitive perception of the environment, which is built up in our everyday behaviour. 'Normal everyday behavior such as a journey to work, a trip to a recreation area, or giving directions to a lost stranger would all be impossible without some form of cognitive map' (Downs and Stea 1973: 10). Cognitive perception and mapping is a result of what de Certeau (1984) defines as territorialisation through spatial tactics. These are daily physical repetitive practices of walking, by which a person's mental perception is constructed. Walking is indeed the most intimate connection of the body to the environment (Madanipour 1996). We react through our body physics to the landscape characteristics that surround us. If we go up a hill we sweat and breathe deeply, if we see beautiful scenery or buildings we get excited. We react to smells, obstacles, pollution, noise, etc. When we walk our body is in a direct contact with what surrounds us and therefore the process of remembering and interpretation is straightforward. Cycling and driving come next and then being a passenger in a car or on public transport. The latter find it more difficult to remember and draw a clear map of their environment (Madanipour 1996).

The construction of mental mapping is a matter of identity. It 'var[ies] from group to group and individual to individual, resulting from our biases, prejudices, and personal experiences' (Downs and Stea 1973: 9). Research shows that people's socio-economic status, ethnicity and race, gender and age, length of residence in an area and travel mode within the city have their effects on people's mental perception and mapping of the environment (Madanipour 1996). Research carried out in Rome and Milan shows that middle-class groups identified a much larger number of elements over a much wider spatial extent because of greater access to mobility as a result of greater access to wealth (Carter 1981). In Los Angeles, people of poor neighbourhoods perceive a much narrower city image than do people who live in wealthy areas. Here the notion of ethnicity and race is closely related with socio-economic status (Carter 1981).

Different age groups and different life cycle stages affect different uses of the city and its perception (Carter 1981). Some basic differences are commonly observed between men's and women's mental mapping as well. As a result of their different gendered roles within the household structures it is assumed that women's mental perception would include a more detailed knowledge of the environment near the home while men's perception of space would be of a wider area but with less knowledge of the details (Fenster, forthcoming). In other research (Harrell et al. 2000) on directing way finders with maps men tended to provide more cartographically complete maps then women, though there were no gender differences in the use of landmarks or labelled buildings.

The understanding of the cognitive dimensions of urban spaces as they are perceived by city inhabitants means exploring their mental spaces, that is, how the 'real' or physical spaces are perceived cognitively and mentally by people who use them. Soja (1989) makes this distinction between the physical space of material nature and 'the mental space of cognition and representation'

(p. 120). The three dimensions of space (physical, mental, social) interrelate and overlap and thus the conceptualisation of space necessitates looking at all these dimensions. This brings us back to Lefebvre's work (1992), in which he introduces the concept of social space as the space of social life, of social and spatial practice. As analysed in Chapter 3, Lefebvre deals with what he terms 'the perceived–conceived–lived triad (in spatial terms: spatial practice, representation of space, representational space' (1992: 40). In this triangle the mental space can be related to Lefebvre's *lived space* or *representational space*:

> Redolent with imaginary and symbolic elements, they [representational spaces] have their source in history – in the history of a people as well as in the history of each individual belonging to that people. Ethnologists, anthropologists and psychoanalysts are students of such representational spaces . . . By contrast these experts have no difficulty discerning those aspects of representational spaces which interest them: childhood memories, dreams or uterine images and symbols (holes, passages, labyrinths). Representational space is alive: it speaks. It has an effective kernel or centre: ego, bed, bedroom, dwelling, house; or: square, church, graveyard. It embraces the loci of passion, of action and the lived situations, and thus immediately implies time. (1992: 41–2)

Representational spaces represent belonging. They represent other notions too, which make them 'private' and 'personal'. Madanipour (1996: 3) makes the connection between cognitive mapping, representational spaces and planning clear:

> The search for a meaning of space is a necessary step to take as it is crucial that before moving into the normative realm of design, we explore the realm of the descriptive and analytical, in other words, to understand urban space before attempting to transform it.

Understanding urban spaces engages the deconstruction of the meanings of comfort, belonging and commitment – as they are expressed as 'local embodied knowledge' in people's own mental maps.

Cognitive temporal maps

Most of the above approaches treat cognitive maps literally as 'maps'. That is, as static entities internally representing an image, information or consciousness of the external environment, which like maps in an atlas are just waiting for their carriers to open them and consult them. What happens to these images as time passes or as their carriers move in space? Can we identify a pattern of temporal changes in these images?

By asking people to draw the three temporal maps of their past childhood environment, of their present environment and their utopian vision of their environment we introduce the notion of *time* into people's remembering and memory. We ask them not just to draw their daily routine paths, nodes, landmarks and edges as Lynch (1960) indicated in his seminal work, but to use their multiple layers of memory, remembering and imagination to draw the maps.

In the research each of the interviewees was asked to draw these three temporal maps after the verbal interview had been terminated. After the interviewee finished their drawings we analysed together the symbols of the three elements of comfort, belonging and commitment in each of the temporal maps. Thus both local embodied knowledge of the interviewees together with the professional knowledge of the researcher were involved in the analysis of the maps.

People's reactions to drawing the maps were mixed. Most people did not mind and even enjoyed the process. But some insisted that they did not know how to draw so they were asked to describe verbally their three environments, as Aliza did at the beginning of this part. The Bangladeshi in London and most of the Palestinians in Jerusalem did not want to draw mental maps. Some agreed to describe their environment in words instead. It is interesting to note the reaction of two planners who were interviewed for the research and were asked to draw the maps. One refused to draw the maps, saying he could not draw visual maps at all. The other drew two temporal maps of the past and the present but could not draw the utopian vision of his desired environment because he said he did not have the vision for it! This perhaps shows how planners, who are themselves the professional makers of utopian environments, sometimes find it hard to visualise and concretise their own environment.

In another work (Fenster, forthcoming) I analyse at length the *temporal* differences between the three maps and their meanings. Here the focus is on the interpretations of comfort, belonging and commitment in the three temporal maps and therefore this rather than time would be the focus of the analysis.

The analysis of the symbols of comfort, belonging and commitment in CT maps

The analysis of the symbols of comfort, belonging and commitment is based both on people's own interpretations of their maps and on a set of indicators suggested below. These indicators were developed based on previous research on mental mapping (Fenster, forthcoming) and on the topics that seemed relevant to the analysis of the expressions of the three elements. Some of these indicators are similar to the categories formulated for the narrative analysis, especially the first and the second:

1. *Symbolic and Physical Images*. This relates to whether the drawings have images representing physical universal meanings or symbolic 'personal'

meanings. For example, symbolic meanings could be 'a hole in a fence' as an indication of childhood memory, or 'secret places' significant to the person who draws because of specific memories. The symbolic meanings of the images indicate, as people themselves say, a greater or lesser sense of comfort and belonging to a specific environment.

2. *Images of the 'private' and images of the 'public' and the boundaries between them*: the extent to which the drawings contain images of the 'private' space and the 'public' space and the extent to which the boundaries between them are clear-cut. For example, people choosing to draw as their environment their inner home or even one room in their home or drawing their environment without indicating their home reflects a certain attitude to private/public divides. It is also interesting to note how people distinguish between their home environment and the 'public', whether they draw a fence, a thin or a thick line. This could teach us about the 'content' of these three elements and the category of space they are mostly associated with.

3. *The centrality and relationships between the home and the environment.* This follows the previous indicator but with much more emphasis on the centrality of home, its size in relation to other items in the drawing and the location of 'home' in the drawing. The image of 'home' represents a strong sense of belonging. It represents the 'private' and the intimate. Its relationship with the environment can also become an indicator of its centrality and of one's sense of belonging in the environment.

4. *Symbols representing individual identity or collective identity.* Symbols representing individual identity indicate those items that are part of a person's everyday life: the buildings, shops, trees, public spaces and other items that they see or use in their everyday lives. Symbols representing collective identity are those items that are connected to the collective, for example: a Star of David that appears in the drawings of Ethiopian Jews of their past environment as an indication of their belonging to Judaism (Fenster, forthcoming). These symbols can teach us the extent to which the person associates a sense of belonging or comfort with their daily practices or whether it is also an issue related to collective identity.

These indicators, together with the knowledge of the people themselves, each analysing their own maps, will be the base of the analysis of the CT maps presented below.

CT maps and urban planning

What can we learn from analysing past, present and utopian temporal cognitive maps that is relevant to planning practice?

Let us look first at the literature dealing with mental mapping and urban planning. Aitken *et al.* (1989) suggest that the most adequate planning of the built environment will be that which reflects perceptions, preferences and images of residents and other users about their present and desired environment. Golledge and Stimson (1997) suggest improving the quality of human decision making by providing more knowledge about people's preferences for, perceptions of and attitudes towards different environments. Madanipour (1996) holds the same line of thinking – that urban spaces should be understood before planners change them by using cognitive mapping. A recent work of Halseth and Doddridge (2001) introduces the KIDSMAP, a project targeted to identify ways of collecting information on what interests children in their environment.

Lynch (1960) was the first to introduce the idea that better planning, design and management of environments could be undertaken if they were based on people's mental perceptions of the environment. Lynch advocated in particular the understanding of images of the world in which people perceive their lives and the world in which they would like to live, in order to improve planning. For Lynch cognition was primarily about learning and communicating ideas about the world. Image was not literally a picture in mind but a cognitive structure, where the mental process intervened in the relationships between the body and the external physical and social world. In his research citizens of Boston, Jersey City and Los Angeles were asked to describe images of their physical environment by drawing sketches. The result of this research was a classification of five significant images in the city that affect people's perceptions of their environment. *Paths* such as streets, walkways, canals and railways are movement channels and are predominant in people's images of the city. *Edges* such as shores, edges of development and walls are the boundaries of the area. *Districts* are the sections in the city that are mentally recognised as having some identifiable character. *Nodes* are the focal points in the patterns of development, such as junctions or squares and street corners. *Landmarks* are the physical objects such as buildings, signs, mountains, etc.

Lynch thought that the apparent clarity or 'legibility' of the cityscape was the most prominent character of the city. Cities with these five elements, argues Lynch, were clearly legible, offer more visual pleasure, emotional security and a heightened potential depth and intensity of human experience. Lynch's five elements of urban images were widely used in urban design to construct more 'legible' environments, although his critics see it as yet another attempt to impose imaginary order in urban spaces from a narrow physical perspective (Madanipour, 1996).

Indeed the analysis of mental mapping and its use in urban design and planning is wider than was envisaged by Lynch. Here we return to the question posed at the beginning of the section as to what we can learn from past, present and utopian cognitive temporal maps that is relevant to planning practice. Analysis of the three temporal maps is a tool for understanding not only the appearance and location of Lynch's five images but especially to learn how people actively experience their sense of comfort, their sentiments of attachment and belonging and their relationships of commitment to the environment using both the indicators introduced earlier and people's own interpretations. This could help in understanding the spatial meanings of the three elements, which can then be translated into principles of spatial planning that better reflect people's perceptions of a good quality of life in the city.

Cognitive expressions of comfort

I have chosen to begin with a 'verbal–mental map' rather than a pictorial–mental map, perhaps because the way Dana analyses the notions of comfort, belonging and commitment in each of her verbal maps was clear to present and to associate with. Dana, a divorced women of 60, mother of two sons, has lived in Jerusalem since 1962. She did not want to draw the maps. She preferred to 'draw' her mental maps with words. Her verbal maps are shown in Figures 9.1–9.3. The translations are shown in Box 9.1.

I then asked Dana to mark the elements that express comfort, belonging and commitment in her 'verbal–mental maps'. She drew round shapes around the words, which represent each of the concepts, and wrote at the side of the page which element she refers to. What she wrote is shown in Box 9.2.

Clearly Dana mentioned comfort, belonging and commitment in each of her CT verbal maps. As will be noted later, for most people the separation between the three concepts was not so evident. A sense of comfort is very much associated with a sense of belonging and a sense of commitment. Comfort is a prerequisite to a sense of belonging, let alone a feeling of commitment. Therefore in Dana's own analysis of her mental maps these concepts sometimes appear together. For Dana it seems easier to define 'discomfort' than comfort, and when she does mention comfort it usually comes together with a sense of belonging.

Dana's 'mental–verbal maps' are all about 'the home'. It is the central space in her cognitive mapping. She expresses feelings of comfort and discomfort at 'home' and outside home but home is always the base. This is true of all her temporal environments. Her mental perception is more related to the 'private' as a point of reference from which her contacts with the outside world are made.

Her childhood verbal–mental map is characterised with a sense of discomfort more than comfort. This sense of discomfort has both physical, absolute meaning – 'the cold house' – but mainly social symbolic meanings – what seem strict social rules for a girl growing up in a religious and probably restrictive atmosphere, 'no boys' and in fact 'no true friendship with girls' are some of the expressions of discomfort in her contacts with the outside world. In her present and utopian verbal–mental maps her sense of comfort or discomfort becomes more physical and more related to the design of her flat – lack of balconies in

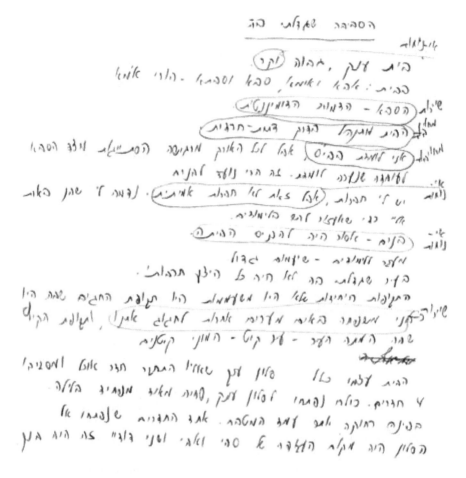

Figure 9.1   Dana – CT map of the past

her flat, and the desire for glass walls and a toilet and bathroom attached to her bedroom in her utopian flat. Again, sense of comfort here is merely expressed at the 'home' level, the very private levels of the six identified categories.

Dana's verbal CT maps show the physical, social and symbolic expressions of the three elements in a very clear way. Her private space – the home – is the focus of her cognitive mapping. The boundaries between the private and the public are clear-cut and the effects of the 'home restrictions' on her contacts with the outside world are very obvious. Her verbal drawings reflect the dominance of her sense of individuality but also the need to be in contact with 'the community', friends and neighbours.

A sense of comfort at an even more intimate space is expressed in Elizabeth's mental map of her present environment (see Figure 9.4). She is single, in her 50s, and has been living in Jerusalem for the last 30 years. She first drew her

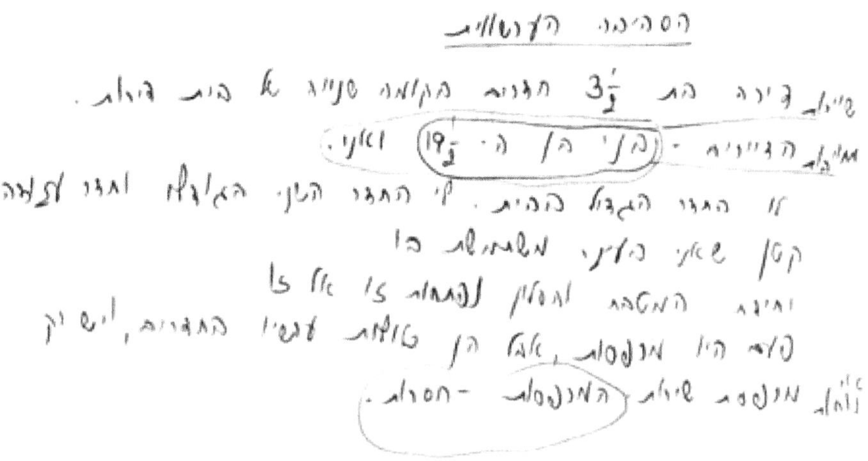

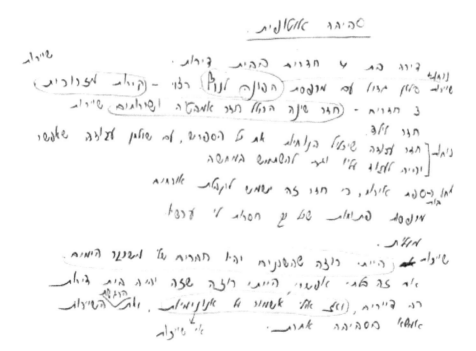

Figure 9.2   Dana – CT map of the present

Figure 9.3   Dana – CT map of the future

BOX 9.1

(Dana) The environment I grew up in – the past

Huge house, tall, cold
In the house: father, mother, grandfather and grandmother – my mother's parents
Grandfather is the dominant character
The household has been maintained as religious – ultra-orthodox
I study in school but all the time I feel the objection of my grandfather to the fact that a girl studies. Studies are only for boys (he thinks)
I have girlfriends but this is not true friendship. I think they come to me so that I will help them at school
Boys – it was forbidden to have them in the house
Despite school – a big boredom
In the town I grew up in there were no cultural activities
The only periods that were not boring were during the holidays when
all the family members from other cities used to come to celebrate with us, the summer was also fun because the town became crowded with local tourists
The house itself consisted of a huge living room that was connected to the dining room and around it were four rooms. All the rooms opened up on the huge living room, which was very scary at night. The kitchen was in the far corner. One of the rooms that opened on the living room was the workplace of my grandfather, my father and my two uncles. It was a bank

Present environment

A flat of 3.5 rooms, on the second floor in a building
The tenants – my son who is 19½ years old and myself
He has the biggest room in the house. I have the second largest room and a small work room that I mainly use
The kitchen and living room are open to each other
There used to be balconies but they are parts of the rooms and there is only one service balcony. Balconies are missing

Utopian environment

A flat of four rooms in an apartment building
Big living room with balcony facing the view. Preferably glass walls
Three rooms – bedrooms with toilet and bathroom
A room for the boy
A work room, which would have enough space for all the books, with a table large enough to work on and have a computer on
Sofa bed, as this room will be the guest room
Open balconies that I miss so much now
Lift
I would like my close friends to be my neighbours and if this is not possible I would like it to be a large apartment building so that I can keep my anonymity. And the feeling of belonging I would have in another space

BOX 9.2

(Dana) The environment I grew up in:

Huge house, tall, cold   → *Discomfort*
In the house: father, mother, grandfather and grandmother – my mother's parents
Grandfather is the dominant character   → *Belonging*

The household has been maintained as religious – ultra-orthodox   → *Commitment*

→ *Commitment*
I study in school but all the time I feel the objection of my grandfather to the fact that a girl studies. Studies are only for boys (he thinks)

I have girlfriends but this is not true friendship. I think they come to me so that I will help them at school   → *Discomfort*

Boys – it was forbidden to have them in the house   → *Discomfort*

Despite school – a big boredom
In the town I grew up in there were no cultural activities   *Belonging*

The only periods that were not boring were during the holidays when all the family members from other cities used to come to celebrate with us; the summer was also fun because the town became crowded with local tourists
The house itself consisted of a huge living room that was connected to the dining room and around it were four rooms. All the rooms opened up on the huge living room, which was very scary at night. The kitchen was in the far corner. One of the rooms that opened on the living room was the workplace of my grandfather, my father and my two uncles. It was a bank

Present environment

A flat of 3.5 rooms, on the second floor in a building
                                             → *Belonging*
The tenants – my son who is 19½ years old and myself   *Commitment*
He has the biggest room in the house. I have the second langest room and a small work room that I mainly use
The kitchen and living room are open to each other
There used to be balconies but they are parts of the rooms and there is only one service balcony

Balconies are missing   → *Discomfort*

Utopian environment

A flat of four rooms in an apartment building
Big living room with balcony facing the view. Preferably glass walls   *Comfort + belonging*

Three rooms – bedrooms with toilet and bathroom   → *Comfort*
A room for the boy
A work room, which would have enough space for all the books, with a table large enough to work on and have a computer on
Sofa bed, as this room will be the guest room
Open balconies that I miss so much now
Lift   *Belonging*
I would like my close friends to be my neighbours and if this is not possible I would like it to be a large apartment building so that I can keep my anonymity. And the feeling of belonging I would have in another space

       Lack of belonging

Figure 9.4   Elizabeth – CT map of the present

three CT maps and afterwards analysed the symbols of comfort, belonging and commitment in each of them. She identified the symbolism of comfort in her bedroom, which she drew as her present environment. She explained why:

> '*I feel comfortable when I sit in my bed every morning . . . this is my room and there is no room for anybody else . . . it is very much mine and I am not willing to move anything or to share it. The table is like an altar . . . it is very private, very much about a feeling of belonging.*'

Here Elizabeth expresses a sense of comfort in a more intimate space than a flat – in her bedroom. In her analysis she herself connects between her perceptions of comfort and those of belonging. As others did previously (Chapters 5–7) she mentions comfort first and then belonging. The sense of comfort comes together with her feelings of attachment. This is an example where physical elements reflect comfort and belonging and perhaps also reflect the very private space that can express comfort and belonging – the bed. There are no signs of contacts with the outside – the public world. It is a very individualistic drawing, emphasising the connotation of comfort and belonging with intimate private spaces.

Fatma, in her drawing of her present mental map (see Figure 9.5) also relates to one room in her house (see the discussion in Chapter 5). For her, her room is her world; Fatma described home as opposed to the city:

Figure 9.5 Fatma – CT map of the present

'Home – prison! Although in my room I have all I need to get "out" – computer, Internet, video, TV cables of 50 channels . . . I have everything but this is not it. City – freedom, personal freedom, atmosphere, spring.'

Being in her 40s and not married, the cultural norms of her family and society force her to live with her mother, where she feels quite suffocated (see Chapters 5 and 8). These feelings of suffocation or discomfort are clearly expressed in her map. See how much detail she drew in her room – with 'everything' she had in it – whereas the rest of the house is more anonymous, as if she is not part of 'her' symbolic home, as she indeed mentions. The contrast between the detailed room and the vague and undetailed home reflects the sharp boundaries between her 'private' and her 'public' spaces within the home itself. For her this is not only the centrality of home and its relationships with the environment but also the centrality of her room and its relationship to the rest of the house. This drawing reflects what she said in the interview when I asked her 'What makes you feel comfort at home?'. She actually reflected on what makes her feel discomfort (see page 125).

Her home is actually her room, where she feels comfortable and that she belongs; the rest of the house is the 'public', it is not 'her' place, it is not her private space and thus she did not draw it on her mental map. Her maps reflect

Figure 9.6   Robert – CT map of the present

only the home, they do not hint of any contact with the outside world; they reflect the individual more then the community.

Robert, who is in his 30s, single, living in London, also drew one room of his flat as an indication of his present environment (see Figure 9.6), but when he analysed his maps he reflected different feelings. His living room is where he feels 'warmth, having my possessions, knowing where everything is, colours, memories'. Both Fatma and Robert drew their rooms as their environments at present. But the differences between the two drawings are very clear. Fatma's room seems as if it is the only space that is hers and it reflects a sense of closure, whereas Robert's living room reflects space, light and openness.

The above examples reflect the centrality of the 'private', the home or the room in people's cognitive mapping of 'their environment'. The actual choice of drawing the home or the room reflects their attachment and probably sense of comfort at home. The home and the room are what symbolise for them comfort and in some cases it is also what brings a sense of attachment and belonging. For them the boundaries between the private and the public are very clear-cut. Their drawings contain icons of the private only, from the bed to the home. Dana is the only one of the four to mention 'the community' in the drawings.

Rebecca expresses a sense of discomfort on a different scale. Using the indicators determined at the beginning of the chapter we can see in Figure 9.7

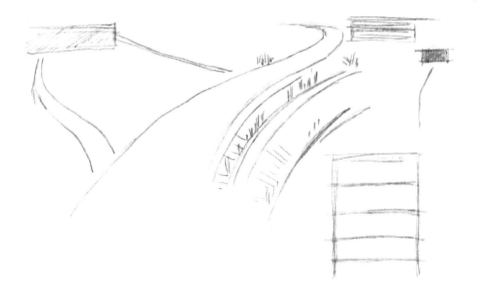

Figure 9.7 Rebecca – CT map of the past

that her point of reference is her neighbourhood, or at least the area close to her home. This is her childhood map. She does not actually relate to her inner home in either map but she does mark her home within the context of her public 'environment'. The tall building in the bottom right corner of the drawing represents her home, with no specific mention of where exactly her home is in this building. There is hardly any connection, no paths or alleys between the home and its environment. It seems as if the home is quite isolated in the area. It is located aside from the main path she drew. It is also hard to relate to the 'private–public' dichotomy in her drawing except that the tall building contains her 'home' and the rest is 'public'. This pattern changes in her present drawing (Figure 9.8), where her home is located in the centre of the drawing and is more connected to the environment, with roads and paths. Both her drawings represent her daily use of spaces: buildings, shops, her work. When I asked her to refer to elements of comfort, belonging and commitment she said:

> 'The house I lived in from the age of 6 to 18 . . . was very tall and I didn't like it so much then and today. The inner flat is not so important and therefore I drew only the tall building. How far it was from everything else. All my first years this was my world.'

The building's height is a physical expression of a sense of discomfort for Rebecca and also its isolation from other elements in the environment. This feeling of discomfort is very much visualised in her past mental map. The distance between the elements and the height of the building she lives in emphasises the sense of isolation she probably felt. Here lack of comfort is physical

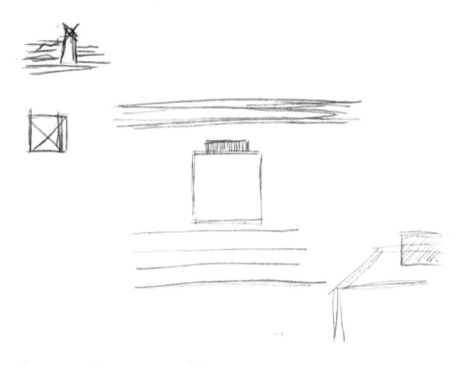

Figure 9.8 Rebecca – CT map of the present

and probably social too. Her mental maps indicate what she terms: 'the boundaries of everyday life' that she herself mentioned when she analysed her maps.

In her map of her present environment she indicates that she drew her home in the middle of the page, which means as she noted that 'home' is more meaningful in her life now than in her childhood. It seems that the location and size of her home in her present map (see Figure 9.8) indicates an increasing sense of comfort: 'The most important thing is the balcony with the view' and she also drew other elements significant in her everyday life: the 'France square' and 'the supermarket' she uses to shop, her place of work and her sister's house with the roof. Her drawing also indicates the two main roads located in front and at the back of her home. It seems, and she also indicated, that she feels more comfortable in her present environment then in her childhood environment. This is reflected in the different location of her home and the better connection between the elements in her present drawing as compared to her past environment.

To summarise, Rebecca's past and present drawings reflect notions of 'private' and 'public' in her maps and also the expression of the relationships between these two spheres. Her home is but a part of a wider environment and the boundaries between the two spheres are quite clear, although not so distinct as in previous examples. Her maps reflect both her inner–personal world but also the contacts she has with 'the community'.

Figure 9.9   Eitan – CT map of the past

Eitan's CT maps provide other insights as to the visual expressions of comfort and especially discomfort in the environment. This sense of comfort or discomfort becomes more explicit when comparing his past and present CT maps. Eitan is in his late 30s, married, living in Jerusalem for some 20 years. He started to draw his past mental map (see Figure 9.9) with the building where his flat/home was located but without mentioning the flat itself. He then felt that the scale of the building was too big and that he would not be able to draw the whole environment so he decided to draw a sketch of his environment in the upper right corner of the drawing. In this sketch he mentions 'home' where the building is located. The home is located on the side of the sketch. He also mentions 'kindergarten' and 'supermarket' as other elements in the environment, which were probably significant in his childhood everyday life. The 'home' is drawn twice. Once on its own in the lower left corner of the drawing and the second time as part of 'the environment' in a map-like sketch. In the first drawing of his home/building it stands on its own with no connection to the environment. In the sketch there is more connection between the home and the environment. These relationships between the 'home' and 'the environment' or the 'private' and the 'public' are expressed differently in the CT map of his present environment (Figure 9.10). The 'individual' and 'community' elements are more connected. His own house is marked with a stronger emphasis,

Figure 9.10   Eitan – CT map of the present

and clear boundaries between the home and the environment are shown by emphasising the entrance to the building and by marking clearly the boulevard of trees that exist in front of his house. He explains what are the important elements in creating a sense of comfort and belonging:

> 'I feel more comfort and identification at present . . . this is not expressed at all in my childhood. I think that comfort and belonging are related to the scale of the city. In childhood the scale of the city was small and my house environment was disgusting, in fact there was no environment . . . and it is expressed physically, the house in the past stands alone, in my present drawing there is a reference between the house and the street, what I miss at present is the lack of connection between my home and the city, there is a connection between the home and the street which becomes part of the house. It is ideal in my utopian map where the house has a strong reference to the city.'

For Eitan, a strong sense of comfort and belonging means a smooth and clear connection between the 'private', his home and the 'public', the surroundings. Indeed his utopian map (see Figure 9.11) illustrates very beautifully what is for him the ultimate sense of comfort in the city. His utopian map contains symbolic images of the city with its paths and alleys and also images of what symbolises 'nature' – trees and greenery. The image of the city takes most of the space in his drawing. It is the combination of living in a home surrounded by a green area and trees that illustrates the quiet and relaxed lifestyle he aspires to with the strong and direct contact with the city and its vibrations, noise and diversity. This combination of 'private' and 'public' and a clear distinction between them is what represents for him comfort in the city.

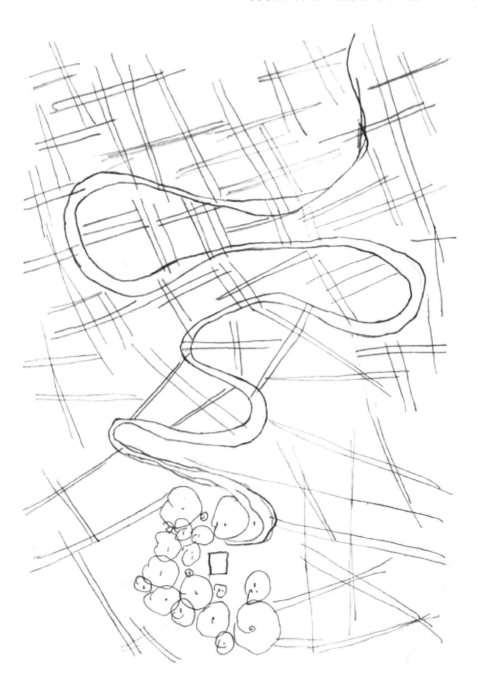

Figure 9.11   Eitan – CT map of the future

This same sense of comfort is expressed in Aaron's mental maps. He is in his 30s, living now with his male partner in Tel Aviv but formerly he lived in Jerusalem. He says:

'What is common to all the maps is the availability of everything. In my past and present maps services are relatively far but still within the boundaries of comfort. In the utopian map it is different, [I hope to live] outside the city, there, there is a need for comfort which is the availability of something that is not intensive, more relaxed, that has more privacy.'

Aaron mentions in his utopian map (Figure 9.12) three words that illustrate the three major components of comfort: 'city', which is marked in the middle of what he perceives as a representation of the city; 'home', which is located on the margins of the city; and 'outside the city', which is illustrated by the trees and the greenery surrounding the city. If we analyse Aaron's map using the indicators that were mentioned at the beginning of the chapter we can see that most of Aaron's images are symbolic. He drew a maze-like image to symbolise 'city', images of trees and greenery to symbolise 'outside the city' and a square image to symbolise 'home', which he surrounded with a dark, thick line – a clear and explicit boundary between 'the home', the private and 'the city', the public. The image of his home is in the middle of the drawing. It is a relatively small image but it attracts the eye as the most realistic image in the drawing, perhaps because of the dark, thick line that surrounds it. It seems that the home is an important place for Aaron and that the clear separation between home and outside home is very important to him too. For him this is the ideal framework of comfort, the proximity and availability of urban elements together with a sense of relaxation that characterise living outside the city. His sense of comfort is very explicit, very physical. It signifies the location of his home in relation to other components in space but with a clear and distinct boundary between the two spaces.

These two CT maps of Eitan and Aaron visually illustrate the combination of the elements that might create a sense of comfort for certain people, the almost contrasting desires of living in peaceful green areas but being in touch, in contact with the vibrant rhythm of the city.

Gender differences and CT maps

Let us look at CT maps of the present environment of Rebecca and Eitan (Figures 9.8 and 9.10). The two are married to each other and live in the same house and still each has a different mental mapping of their environment that reflects the different everyday life experiences and spatial movements they undertake. Eitan, who works at home most of the day, emphasises home in his drawing and also drew the nice boulevard of trees, which is located in front of their home. His drawing refers to the environment such as neighbouring buildings and streets. Rebecca's workplace is outside her home and therefore the

Figure 9.12 Aaron – CT map of the future

space she related to in her mapping is larger. She also mentioned elements that are connected to her family contacts such as her sister's home. Rebecca mentioned elements that are relevant to her domestic duties such as 'the supermarket', although from conversations with them they say they share the housekeeping tasks equally. It is interesting to note that Rebecca did not mention the boulevard in front of their home, perhaps because she does not use it quite so often, as sometimes she feels intimidated by those who sit on the benches in the boulevard. When she talked in the interview about 'what makes her feel comfort in her street' she said:

> 'What is not comfortable in my street is the boulevard, which I don't use because it has only one exit in the middle and there are benches along it and strange people sit there and I feel "locked in", especially at night when the boulevard is full of homeless people and it's not pleasant to pass it. If you get into it you're "locked" in it. It is a planning mistake.'

For Rebecca this space is dangerous, especially at night, and as a result she does not use it at all and perhaps that is why she had not shown it in her drawing. We have already mentioned this narrative in Chapter 8 when we discussed the role of urban design and planning in creating spaces of fear, become as the boulevard has become for Rebecca. Besides this point, there are no significant differences between Rebecca's and Eitan's maps that might have indicated how gender identities affect different perceptions of space. It must be emphasised that the two do not have children so that the household duties do not include the role of child rearing – usually the woman's roles – which dictate a different use and perception of space.

The right to comfort in the city as expressed in CT maps

'The right to comfort in the city', a concept that was developed in Chapters 3 and 5 as part of 'citizenship in the global city', is clearly illustrated in Saida's CT maps. Remember, 'citizenship in the global city' and 'the right to comfort in the city' are closely related to the right to maintain and respect difference. This is connected to the right to belong, that is, if one's difference is recognised and respected one feels one belongs, or if cultural norms and values of communities are respected their members feel more committed. 'Citizenship in a global city' includes the duties of states and cities to their citizens to design and manage 'the city of memory, of desire, of spirit' but also a city that reflects people's perceptions of comfort.

Saida's maps are verbal. She mentioned key words that describe her three temporal environments and then marked which concept (comfort, belonging, commitment) represents each element (see Figures 9.13–9.15). In her three temporal maps the notion of comfort and discomfort (marked as C – Comfort and DC – Discomfort) are the most dominant. She describes comfort and discomfort both in her 'private' space – the home – and the public spaces she

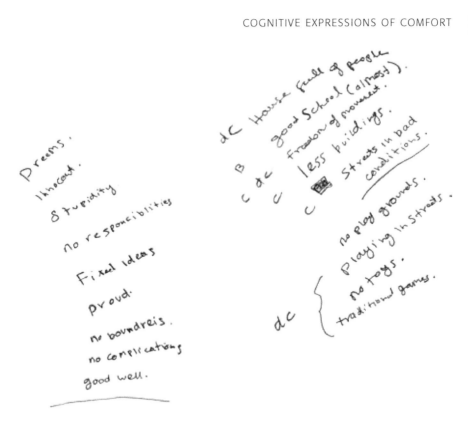

Figure 9.13   Saida – CT map of the past

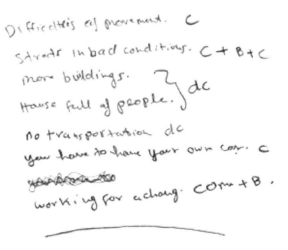

Figure 9.14   Saida – CT map of the present

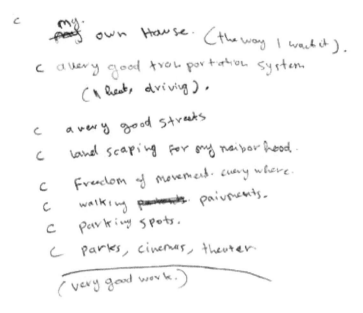

c   my own House. (the way I want it).

c a very good tron portation system
(& heat, driving).

c   a very good streets

c   land scaping for my naibor hood.

c   Freedom of movement. every where.

c   walking ~~pavements~~. paivments.

c   parking spots.

c   parks, cinemas, theater

(very good work.)

Figure 9.15   Saida – CT map of the future

lives in – East Jerusalem and her village, which is located near Jerusalem. It seems that most of what she relates to as 'Discomfort' are those elements in her everyday life that are missing because of inequalities in development, infrastructure and services provision between the East side and the West side of Jerusalem, elements that are so lacking for Palestinian residents of Jerusalem. We can thus see her 'map' as an expression of abuse of her 'rights to comfort'. Indeed if we look carefully most of the elements in her verbal maps that deal with comfort and discomfort are physical and have direct relevance to urban planning and management: 'streets in bad condition', 'more buildings', 'no transportation' are only part of a checklist of the physical expressions of the abuse of her 'right to comfort' in the city. She also mentioned as part of discomfort 'difficulties of movement' that are linked to the times of the second Intifada', when roads were blocked by Israeli military forces (see the analysis in Chapter 4). Her utopian map is actually a checklist of elements that are all connected to what comfort means. As we can see she relates both to private issues such as 'my own house' adding in brackets '(the way I want it)' or 'very good work', which she added at the end of the list in brackets as well. But the rest of the list engages 'public matters' concerning lack of development in the Palestinian side of Jerusalem. Her comfort in the city includes 'very good transportation system', 'very good streets', 'landscaping for my neighbourhood', 'pavements', 'parking spots', 'parks, cinemas, theatre'.

To sum up, comfort is sometimes expressed in CT maps in its negative sense, that of discomfort, especially for those who describe childhood experiences.

Comfort has its expression in both the physical and the social elements. A sense of comfort is reflected in the different scales of space, from the very private – the room – to the more public, engaging both 'the home' and 'the environment' in the various drawings. Comfort thus is identified by the relationship between 'the home' and 'the environment', usually the city. It is connected to availability of services and to the intricacies of living in the city but enjoying the tranquillity of the suburbs.

Comfort is obviously an identity issue both at the individual and community level, especially when it touches the notion of 'comfort in the city', which engages issues of equality and difference. The expressions of 'the right to comfort' become part of the discourse around citizenship in the global city, incorporating equality in planning, development and city management.

A sense of belonging and CT maps

The creation of the schemata that Lee (1968) defined earlier, which is actually the basis of our cognitive mapping of our environment, is built up on repetitive activities in our daily lives. These repetitions, which establish the schemata of our environment in our brain, also construct the sense of belonging, the everyday belonging to our environment. The two notions are consequently connected. The process of the schemata creation in our mind is also the process in which our sense of attachment and belonging to a place are determined.

> '*My sense of belonging [to my past environment] is hierarchical – to the street, to the friends who lived there, to the block of buildings, to the passages between the buildings through backyards, and then the neighbourhood, the garden, the grocery . . . and then the bigger neighbourhood, which includes school and what is beyond it. These elements are not included in the map.*' (Aaron, 30s, single)

This is what Aaron described after he drew his three CT maps. Aaron connects between a sense of belonging and memories of childhood. What he reflects is an intimate, personal and private knowledge of his childhood environment. Such an intimate knowledge is usually a result of his repetitive walking practices. Aaron connects in his past map memories that are both spatial and social or symbolic (see Figure 9.16). He first drew 'the physical layout' and then he filled it with his personal and intimate social content, relevant only to him – his private environment.

'*Belonging refers to people*' says Liat when she refers to expressions of belonging in her past map (Figure 9.17). Indeed her past map is a mixture of physical images and symbolic meanings. In drawing her past map she made efforts to be as precise as possible about the physical layout and the landscape of the street she lived in. In addition, she mentions the type of trees that existed in her street: 'Fig tree', 'Margosa tree', 'Vines', etc. After she finished drawing

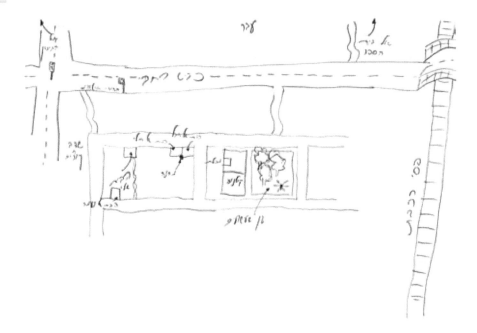

Figure 9.16   Aaron – CT map of the past

the physical layout she labelled each house with its symbolic social meaning for her. She mentions the names of the neighbours with personal reference to each of them: 'Family Azar – the best neighbours', 'Itzik Alfasi – a friend two years older then me', 'Edna Cohen – one year older then me'. This is more than just name-dropping; she also mentions the social importance of each person in her childhood. Liat's past map is a spatial expression of belonging, a social belonging. It is an expression of her *representational spaces* using Lefebvre's terminology, which is constructed through images and experiences.

Memory is part of an individual identity as much as it is part of the collective identity. Sites of memory are sometimes intimate and personal, as in the case of Liat and Aaron. They reflect personal memories of everyday life in childhood. Sites of memory also reflect collective memories and identities, 'public memories' that find their ways as sites of commemoration in the city. Both layers of memory, individual and collective, construct a sense of belonging to the city but they do not necessarily find their expressions in people's mental mapping of their city. The Arab–Palestinians who were interviewed in this research did not incorporate in their drawings any symbols of the Palestinian–Arab collective memory and belonging, even though this sentiment had been strongly expressed in their narratives. Their drawings engage their everyday life experiences, which are related to their personal memory more than to their collective memory. They drew their 'private' surroundings like anybody else. In Figure 9.18 we can see one such example: Magda, who is in

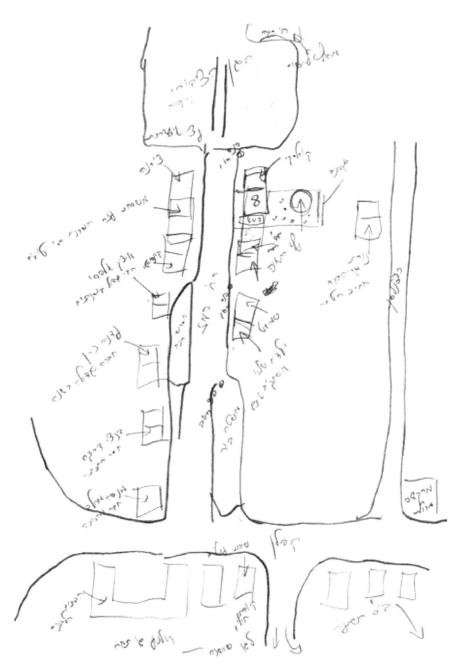

Figure 9.17   Liat – CT map of the past

Figure 9.18 Magda – CT map of the present

her late 30s, is a Palestinian citizen of Israel who works in West – Jewish – Jerusalem but lives in East – Palestinian – Jerusalem. She mapped her every-day life environment without drawing all the elements that make her feel she belongs or dis-belongs to Jerusalem which she mentioned in the interview. Those elements are mostly connected to the abuse of her citizen rights in not having equal levels of services and infrastructure to Jewish citizens. This is probably partly because of her ability to draw, but mostly perhaps because when she was asked to draw her environment she had a cognitive mapping of her everyday practices more than of her identity related collective memory.

CT maps as expressions of nostalgia

'The deepest sense of belonging, my roots to the city, to the home, to the place, the roots of the tree, this is where I came from, here I would feel a stronger sense of belonging [at my home], this is why I drew it, it is a point of reference. The comfort is that I could do all of this . . . There were no buses then, no money, I paid for my own studies but there was a lot of good will. It was a different

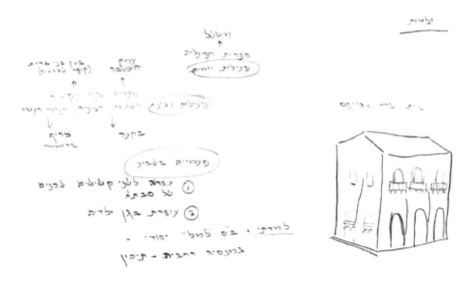

Figure 9.19   Sarit – CT map of the past

*culture, there were no keys and locks to the doors. A lot of community life, espe-
cially in my childhood area, I felt connected to everyone, we belonged to each
other.'* (Sarit, 50s, married, living in Jerusalem)

Sarit describes her childhood with many nostalgic feelings. Her home is obviously a central emotional part of her childhood, and indeed she drew it as a
central figure in her past CT map (Figure 9.19). The house seems relatively
isolated in the area and because she felt she did not have enough space on the
page to draw all her other activities she wrote them. All her teenage activities
are mentioned, including the place where these activities happened and the
number of times per week she did them. It is interesting to note how nostalgic
and idealistic is her description of her childhood environment. She describes an
ideal community where locks were not used on doors and this sense of trust is
what made everybody feel they belonged, a mutual responsibility.

Nostalgia is associated with belonging, belonging to family, friends and the
community. Recently nostalgia has reappeared in postmodern geographical
discourses together with notions of image, fantasy making and archetypal
psychology as tools for analysing postmodern texts discussing the relationship
between the 'real' and the 'image' or the fantasy in landscape and other components of everyday life.

Nostalgia for the past can sometimes develop as a result of a lack of belonging in one's present life. It is a search for past places, places that contain frac-
tions of individual identities (Bishop 1992). Thus nostalgia can be interpreted
as a regret for the lost 'self', a self that we all lose when moving ahead in
time (as we grow up) and in place (as we move places). Sarit's narrative is a

reflection of a loss of a sense of community and security that she longs for. Nostalgia is not merely a matter of regret for lost time, it is also a pining for lost places, for places we have once been in and can no longer re-enter (Casey 1993). Taking this view, we are actually all displaced people in many ways (Casey 1993), as we all become nomads, migrating across a system that is too vast to be our own but in which we are fully involved, transcending and trans-forming bits and elements into local instances of sense. Nostalgia thus is a speaking symptom of the profound placelessness of our times, in which we have exchanged place for a mess of spatial and temporal pottage (Casey 1993: 37).

Time is actually what makes perceptions of space different for the same person. It is time that creates memories and nostalgia. Memories of the past are usually spatial and place connected. It is usually when somebody speaks of the past that their points of direction are connected with places and spaces. The 'art of memory' that existed prior to the eighteenth century was concerned with placing and organising images in relation to each other. Memory at that time was perceived as a quality of discrete interior space. The process of remember-ing then becomes a journey through this 'interior' landscape to reclaim these images (Bishop 1992). Defining nostalgia in terms of time actually means an attempt to connect the two contradictory modes of time, the objectified, socially measurable time vis-à-vis the personal, subjective, imaginative time created by the 'self'. The personal/subjective environment – the past environ-ment is a biography in the sense that it is about a subjective oral or literary scheme or model telling about an individual's life (Beckett 1996). Its subject-ivity is reflected in the fact that nostalgia usually creates a 'place', different from what used to be in the past and from its present form, a place whose meaning and cognition is personal and subjective and therefore people perceive the same place differently. Efrat's mental maps (Figures 9.20 and 9.21) reflect this subjective–objective distinction. She is in her 40s, married with two children and identifies herself as Israeli–Tzfonit (from an affluent area in Tel Aviv). She drew her childhood map with a lot of attention. She concentrated very hard and it took her some time to draw it. She made efforts to draw all the many details that appear in the map and to mention their function and symbolism for her. Then she drew her present environment. She did it very quickly. It took her a few seconds to finish the drawing. When we talked about the drawings she said:

'My present map is the opposite of comfort. This highway that I hate so much. There was a lot of comfort in the past. It was an open space in which it was pos-sible to walk and ride a bicycle. The past environment was very diversified. There was a lot of fun then. There is fun now too but its more monotonous, too new, belonging at present is more functional.' (9 August 2000)

Eftrat's past map symbolises her nostalgic feelings for her childhood. The map contains many details that reflect the diversity of activities she had in her

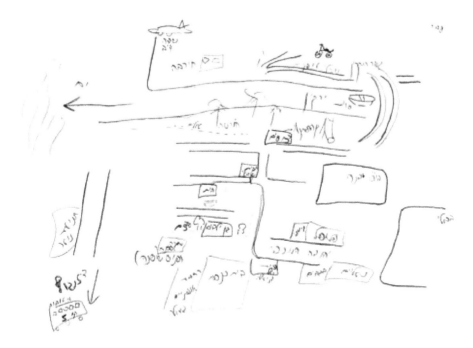

Figure 9.20 Efrat – CT map of the past

Figure 9.21 Efrat – CT map of the present

childhood. Visually it seems like a more open and accessible space with no barriers. The highway in her present map is a big barrier. She started drawing the map with this barrier. She herself mentions that her present map is more utilitarian whereas her past map is more subjective–personal, where she draws her personal experiences together with the physical environment. The differences between the two maps reflect her nostalgic feelings and although she talks about comfort more than belonging it is a narrative about a sense of memory and belonging to her childhood environment.

Elisabeth Wilson in her paper '*Looking backward: nostalgia and the city*' (1997) writes:

> To return to a city in which you used to live is – especially if the gap is a long one – to be made sharply aware of the passage of time, and the changing fabric of cities congeals that process of the passage of time in a way that is both concrete and somehow eerie and ghostly. It reminds you that the city is neither of those two favourite figures of speech: a work of art or a diseased organism. It is a process, a unique, ongoing time/space event. (p. 128)

The city is actually neither 'utilitarian/objective' nor 'personal/subjective' but a mixture of the two, creating the personal identity of a place. Efrat's drawing of the past can be perceived as a *representational space*, a space that she perceives through images and symbolism from the point of view of today, of the present. It is representational space but from a viewpoint of a different time that makes it more nostalgic.

The centrality of home as a space of belonging

The nostalgic explanation of the centrality of the 'home' focuses on Constable's paintings of the nineteenth century. The cottage in these paintings was the root metaphor not just of the English home but also of homeliness. The home is like a doorway to the psyche argues Bishop (1995). He imagines the home as 'an entrance to ancestral underworld'. Here the centrality of home is explained as almost a psychological necessity for stability and connections with past roots and this provides a partial explanation for its centrality in some of the cognitive mappings.

In Rebecca's drawings (Figures 9.7 and 9.8) we saw how the image of the home is more central in the drawing of the present than in that of the past. That is, in the map of the present, home is central as compared to its more marginal location in the past map. Mary's past and present maps (Figures 9.22 and 9.23) reflect the opposite. Mary is in her 40s, single, and identifies herself as Israeli – Ashkenazi–Jewish. She lives in Jerusalem. In her past map the home is located in the building, itself in the centre of the map. The home is divided internally into its rooms and family members. There is a balance in the detailed

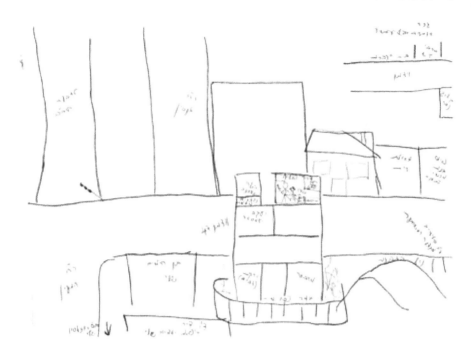

Figure 9.22   Mary – CT map of the past

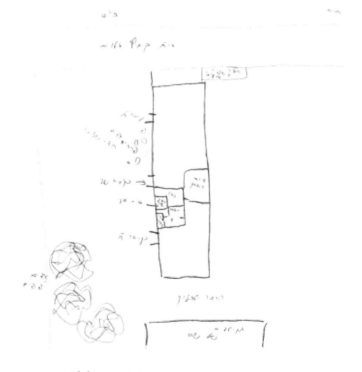

Figure 9.23   Mary – CT map of the present

information between the home and the environment, that is, the environment becomes part of the home. In contrast, in her present map her home seems less central, although she does mention it and its internal division.

Eleonore's past and present maps reflect the same character (Figures 9.24 and 9.25). Eleonore is in her 50s, she is single, and identifies herself as British – white. She lives in Britain. In her past map 'my home' is put at a road junction, which attracts the eye as the focal point of the map. The size of her home is similar to the size of other elements in her drawing, but it is the location of the home that makes it more visible and central. In her present map 'home' is mentioned somewhere along the relatively empty map, which is less detailed than the past map. Eleonore placed much emphasis on her home in the interview. She emphasised the importance of her parents' home as part of her sense of belonging while she felt that she belonged less and was less attached to the area she lives in at present. The different locations of 'home' in her two temporal cognitive maps illustrate this difference.

A different view of home appears in Lynne's drawings (Figures 9.26 and 9.27). She is in her 40s, single, and identifies herself as Scottish – white; she lives in London. Both her past and present cognitive maps consist of her home only! In her past map home contains her family members similar to the way in which Mary drew her past map. Her present cognitive map consists of only one element, 'refuge', which is the symbol and the meaning of home for her. Both her

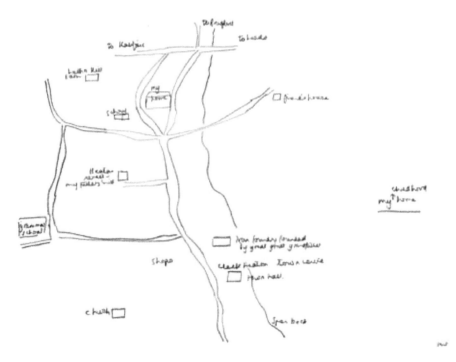

Figure 9.24   Eleonore – CT map of the past

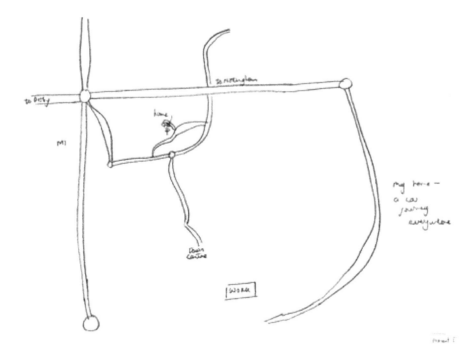

Figure 9.25   Eleonore – CT map of the present

Figure 9.26   Lynne – CT map of the past

Figure 9.27   Lynne – CT map of the present

temporal maps express emotional feelings towards the notion home, reflecting her sense of belonging and emotional attachment to the meaning of home in her life. In Chapter 6 we mentioned Massey's (1994) challenges of the notion 'home' as having multiple meanings and multiple places. The centrality of home in the temporal cognitive maps does not necessarily reflect the physical home but the sense and the meaning of home, subjective in itself for these people who drew their maps. It probably signifies the importance of stability and connections with social and physiological elements in each individual's life.

Belonging and possessions

Possessions, their existence and their order were some of the elements people mentioned in the interviews as making them feel they belonged to their home. Two illustrations appear in both Robert's and Elizabeth's present cognitive mappings (see Figures 9.4 and 9.6). Their cognitive mapping of their present environment was one room in their flat. For Elizabeth it was her bedroom and for Robert it was his living room. In his interview when asked 'What makes you feel you belong to your home?' he said: 'My possessions and how I set them out.' The space in the home where his possessions are set out in a way that makes him feel a sense of belonging is his living room. Elizabeth drew her bedroom and when we analysed her drawings together she said: 'this is my room. It's very much mine, it's very personal, it's very much about belonging'. It is not only the home as a whole that is associated with a sense of belonging and attachment, it is a room in the house that is more significant than other parts in making the tenant feel belonging and attachment and it is the posses-sions in that specific room that create this sense of belonging. The same as with the cognitive expressions of comfort so is it with belonging – the intimate, per-sonal space is sometimes the space most reflecting this feeling.

To sum up, a sense of belonging as expressed in the cognitive temporal maps is more associated with the private, the intimate space within the home and less with 'public' or 'communal' issues such as national identity, notions of exclusion and civil rights as they were clearly expressed in the interviews or even in verbal maps (see for example Saida's cognitive maps, Figures 9.13–9.15). Belonging is associated with the physical elements in one's environment but also with the social and symbolic meanings of these elements and as such consists of the emotional mappings of one's own space and environment.

The cognitive symbols of commitment in mental maps

'*Commitment is the future . . . it is something personal which expresses commitment to Jerusalem, a will to "fix" or change Jerusalem . . . The ideal city doesn't include my home.*' (Rebecca, 30s, Jerusalem, 3 December 2000)

In her future utopian map Rebecca drew Jerusalem as a city with an imagined river (Figure 9.28). This is because as she says it seems to her 'not natural' for a big city such as Jerusalem to function without a river, as most cities in Europe are located near rivers and rivers serve as a point of location, of reference.

Figure 9.28 Rebecca – CT map of the future

Rebecca reflects notions of commitment only to her utopian cognitive mapping. She says: 'it is something personal . . . a will to "fix" [the city] . . . the ideal city doesn't include my home'. Why does the ideal city not include her home? She does not specify, but this is a statement of non-commitment, a commitment to change things in the city but not to be part of the city.

As much as Rebecca reflects the notion of commitment in her future utopian map so do most of the interviewees. When analysing the symbols of comfort, belonging and commitment in their maps most of them associated the notion of commitment with their utopian imagined future map. Commitment is cognitively associated with future environment, more than with present or past environment. One explanation of the association between commitment and future utopian cognitive maps is that in their future environment people express their desires and aspirations as to the way their environment should be. They feel more willing to commit to their own fantasies because by being thus committed they in fact are committed to themselves.

Mandy drew her utopian environment as including four places where she feels emotionally attached: 'Montreal–Hyderabad–London–Jerusalem' (see Figure 9.29). Her map contains physical elements accompanied by verbal meanings: 'home', 'cinema', 'work', etc. There are also symbols of people in all her environments and a symbol of the 'sun', which appears in the middle of her map. She says: 'sun is about commitment – a commitment to make a place

Figure 9.29   Mandy – CT map of the future

sunny'. This is why the sun is a central figure in her utopian map next to the image of 'home' and 'green spaces'. 'Home' is indicated quite clearly. Home is bigger in relation to other images and contains a very clear boundary, which symbolises the distinction between 'private' and 'public' in her drawing. Her collective identity is expressed very explicitly in one of the sketches, of Hyderabad located in the lower right corner of her drawing. She mentions the word 'community' and also draws figures of people to symbolise the community sense she wishes to have. In the map she also mentions 'no gates'. In this she refers to her desire to break the sharp boundaries between the different sections of the city in Jerusalem. Later on, when she analyses the notion of commitment in her utopian map, she explains: 'commitment is the future, community, no gates, no guns' (16 June 2000). She is a Muslim who lived in the Jewish side of Jerusalem and for her the utopian future is a city with no guns and no gates, the physical and the symbolic gates between West and East Jerusalem.

Susana also associates the notion of commitment to her future utopian cognitive mapping (Figure 9.30). Her utopian map is relatively simple compared to her past and present maps (Figures 9.31 and 9.32). Her map reflects her private desires and aspirations. This is the only map in which her home is so explicit and so big, in contrast to her present and past map, in which home is not mentioned at all. When analysing the maps she said: 'commitment reflects the future: family! A few years ago my commitment was to the revolution [in Czechoslovakia], saving the world. Now it is different although I do care about the city' (13 July 2000). Being one of the leaders of the students protest move-

Figure 9.30   Susana – CT map of the future

Figure 9.31   Susana – CT map of the past

ment in the former regime of Czechoslovakia, Susana was very active out of a deep commitment to her former country, but now she has become a mother she feels more committed to her family, as she mentions both graphically and verbally, in that her utopian map not only indicates a big home but also a clear boundary drawn in a line that marks a separation between her private space and the public, outside this intimate boundary. Her three drawings clearly indicate the multi-layered nature of commitment and its temporal nature. Its content and meanings change depending on one's own life cycle. It changes especially when one becomes a parent. Susana's past and perhaps present life signify commitment to public matters while her desires for the future focus on commitment to her own home.

For Sarit: 'commitment will create comfort [in the future]'. This is what she said when she analysed her utopian map, which is verbal rather than graphic. In the map she literally lists the various activities that she thinks she should be involved in. These activities are for her the components of commitment. She calls her utopian cognitive map: 'my utopian perspective'. She wrote 'me' in a square and from that square drew several arrows to indicate her multiple activities (see Figure 9.33). She writes: 'I carry on with cooperative activities, I carry on with studying environmental studies, I carry on with educating the citizen how to live in a quality environment, etc.' At the end she adds: 'to carry on with

Figure 9.32 Susana – CT map of the present

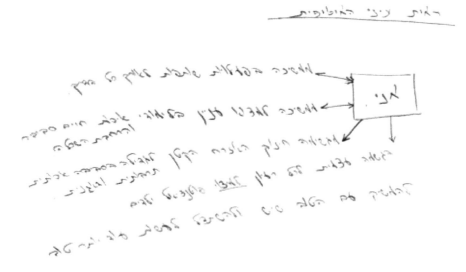

Figure 9.33 Sarit – CT map of the future

the good that there is and to make efforts to make it better'. As she explained later, for her everything she mentioned is the essence of commitment. For Sarit commitment is about 'carrying on', it is about public activities. She is also a mother but her children are grown up and her essence in life is how to help and be involved in communal and public activities.

Although commitment is mostly associated with the future some people did connect this element with the past, although this connection was not reflected in their mental maps of the past. Julian (37, single, Canadian – Israeli–Jewish, Jerusalem, 10 February 2000), for example, said when analysing the notion of commitment in his past and present maps (Figures 9.34 and 9.35): 'past map certainly reflects comfort and belonging: childhood, family. Present map reflects comfort . . . commitment is the Old City, its spirituality'. Here commitment is expressed in reflections on the sense of bonding to the symbolic sites in the Old City, which represents past Jewish history. Commitment is also associated with the notion of belonging and attachment. It is a 'collective' commitment resulting from a collective memory and identity.

Liat mentioned this association between commitment and past history as well, but in her case it is the personal past not the collective, historical past. When she analysed her temporal cognitive map (Figure 9.17) she said: 'commitment is to my childhood, to my life and my memories. At present commitment is to the commemoration of the past and this is what I do as a profession' (31 July 2000). For Liat commitment relates to two temporal spaces and memories: commitment to her personal past, her childhood, which she drew in a very detailed manner, and 'collective commitment', which is similar to what Julian mentioned. This commitment to the collective is articulated in her present occupation. She works in a museum, which commemorates significant past events in the history of the state of Israel. However, unlike her personal commitment to her past, which is illustrated in detail in her past cognitive map, the commitment to the collective that she perceives as engaging in in her present work has no graphic expression in her cognitive map.

The above analysis reflects the commitment to continuity, to the future. It is commitment to be part of sentiments we feel and believe in, commitment to an idea or a project that one wants to be part of. This has a very strong expression in Sarit's narrative in the repetition of the word 'carry on' and also in Liat's narrative.

As mentioned in Chapter 7, commitment is the most complicated element to define, as were its graphic expressions. It is nevertheless interesting to note that those who connected commitment to graphic visualisation did so mostly in connection to their future utopian maps, although in the interviews they talked about actions and activities that are part of their current life. One explanation of this difference could be that perhaps in principle all actions at present that are part of a sense of commitment are targeted to create a better future, and the representation of the future in the utopian map is the best expression or even the end result of this action of commitment.

Figure 9.34 Julian – CT map of the past

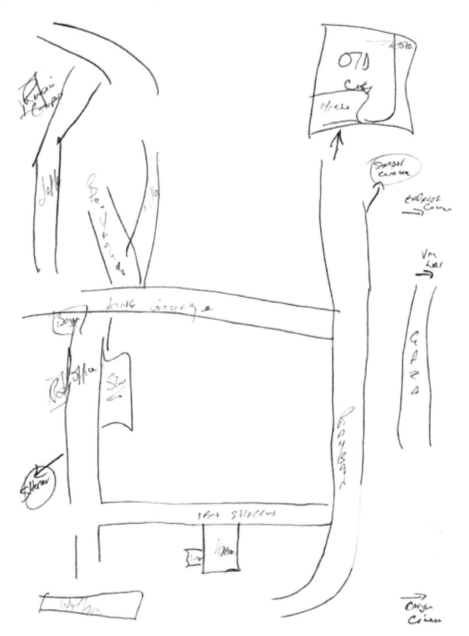

Figure 9.35 Julian – CT map of the present

Summary and conclusions

This chapter had two main goals: first, to establish a framework within the context of cognition and cognitive maps for analysing people's drawings as another means of understanding the notions of comfort, belonging and commitment to their environment. The second goal was to link this framework with urban planning and to suggest this tool as another means of improving the link between people's needs and aspirations and urban planning and management.

Let us summarise the topics raised in the chapter with regard to these two goals:

- *The analysis of the meanings and expressions of comfort, belonging and commitment consists of two main lines of knowledge*: local knowledge reflected in people's own interpretations of their drawings and the professional–academic knowledge expressed in the set of indicators that were formulated at the beginning of the chapter.

- *Comfort, belonging and commitment are time-spaced expressions.* In general it seems that the notion of belonging has stronger connections with the past, with nostalgic events and memories. The notion of comfort is more explicit in people's present mental maps as it usually relates to the physical elements in the environment that have social or symbolic meanings. The notion of commitment is clearly expressed in people's future utopian mental maps as it includes the desires and aspirations of the best fantasy place to live in, a fantasy place that will make its inhabitants committed. These meanings of the three elements have not been expressed as clearly in people's narratives and this is the contribution of cognitive mapping – it adds dimensions and in-depth understanding of these concepts.

- *Palestinians' mental maps reflect their sense of abuse of the right to comfort in the city* in that such maps represent the hardship of everyday life, which mainly results from lack of infrastructure and services.

- *The significance of analysing comfort, belonging and commitment to the planning practice.* Analysing the meanings of comfort, belonging and commitment emphasises the use of CT maps as a means of exploring significant components in people's everyday lives. This analysis identifies the overall cognitive mapping and perception of people in terms of the relationships between physical and symbolic images, the boundaries between the private and the public, with particular reference to the centrality of home and the relationships between their individual and communal identities and spaces. This analysis can provide planners with an in-depth understanding of different aspects of quality of life, an element that is usually defined as one of the targets planning aims to achieve. Other researchers (Halseth and Doddridge 2000) suggest the use of KIDSMAP for identifying the needs and aspirations of the clients of the city, in their case the children who live in the city. In their research they used Lynch's elements to analyse children's maps.

Here we suggest linking CT maps and urban planning in a different manner. We argue that the fact that CT mental mapping evokes different emotions, perceptions and attitudes among people than those raised in the verbal interviews shows that there are certain aspects to quality of life that are expressed in methods other than verbal that planners can incorporate in their work with communities in order to expose the optimum level of local knowledge.

What knowledge can we draw out from CT maps that can assist planners in their work with communities? If we take the indicators presented at the beginning of this part we can draw some links between them and give planners a better understanding of aspects of quality of life:

1. *The symbolic and physical images that are part of the mental maps.* As already mentioned in the analysis, physical and social images could be personal or collective. They could indicate sites of memory and belonging, especially the collective ones. In identifying the needs and desires of community members, such as sites of belonging, they could be incorporated in planning schemes for urban spaces. This point is further elaborated on in Part IV.

2. *Images of the 'private' and the 'public' and the boundaries between them.* These images can help planners understand how residents of a neighbourhood, for example, construct the spaces within their area of living, that is, which area they identify as private and which areas as public. Planners can identify the different expressions of 'private' and 'public' and the extent to which the boundaries between the two spaces are clear-cut. After identifying these divisions planners can interpret these expressions with the community members themselves in order to have a clear ideas as to how to reshape these spaces in ways that perhaps better connect between private and public activities.

3. *The centrality and relationships between the home and the environment can help planners in identifying problems* of accessibility and transport in certain parts of the city. Some of the mental maps indicated the desired relationships between one's own home and the city. The CT maps of both Eitan and Aaron (Figures 9.11 and 9.12) indicated that for them comfort means living at 'home', in this case not only their home but in a more residential area while remaining connected to 'the city' – the vibrant and active spaces that the city represents. The two expressed a clear desire for smooth accessibility between the two major spaces.

4. *Symbols representing individual identity and collective identity* serve as indicators of the images of belonging and attachment in urban spaces. This could serve planners whose task is to design or redesign certain parts of the

city. Sites representing collective identities can become sites of preservation. These sites are sometimes not public enough to be declared national sites of preservation but they can still have some significant meaning to the residents of the area. CT maps can serve as a tool to identify such sites.

Obviously the practicalities of incorporating such knowledge in planning practice depend on the *scale* of the plan. The more detailed the scale of the plan the more meaningful and useful it is to include such local knowledge in the planning process. It is probably of most use when planners intervene at the neighbourhood level but it is also very relevant when planners deal with other parts of the city.

Last but not least, the practicalities of incorporating local knowledge embedded in CT mapping depend on the *power relationships* between the community, the planner and the authorities. Evidently when a plan is initiated by the community members and the planner acts as a facilitator or as a 'translator' of local needs, desires and aspirations the significance of this method is greater and the plan will be a direct result of individuals' and the community's aspirations. The more 'top down' the plan is and the more establishment oriented the planner the less significance such local knowledge has in the planning process.

PART IV

Between the 'Holy' and the 'Global':

On Local Embodied Knowledge and Spatial Planning

This part challenges the validity and viability of the expressions of comfort, belonging and commitment in planning practice. This is the last step in the process that has been developed in the book of formulating a framework of local knowledge. This framework is based on a different understanding of everyday life in the city, a city that is made up of our homes, our buildings, our streets, our neighbourhoods, our city centres and urban parks; a city that is designed, planned and managed by an increasing number of actors and interest groups; a city that is real and imagined; a city that consists of a variety of homes and is a home in itself.

In the final section of this part several case studies will be presented, illustrating the complexities and the opportunities of incorporating the three notions in democratic planning structures in order to feel at home in our city, our home.

CHAPTER 10

Local knowledge and the planning of the built environment – lessons in practice

This chapter challenges the validity and viability of the expressions of comfort, belonging and commitment (CBC) in planning practice. This is the last step in a process of formulating a conceptual framework of local knowledge that has been developed in the book. The framework is based on a different understanding of daily practices in the city, a city that is made out of our homes, our buildings, our streets, our neighbourhoods, our city centres and urban parks. How we can make ourselves at home in the city is perhaps the major challenge of this new framework of local knowledge and its incorporation in urban planning and management.

This framework of understanding of local knowledge began with the analysis of the existing planning traditions, highlighting the sources of knowledge – professional and local – that each represents. It then continued with deconstructing identity issues as reflected in people's everyday experiences in the city, in particular looking at notions of exclusion and inclusion and defining the components of 'citizenship in the global city'. The analysis of the expressions of comfort, belonging and commitment as they relate to the different categories of space was the next stage in this framework. Then the different experiences of CBC of men and women were the focus, clarifying the extent to which gender identity matters. This was followed by an analysis of people's cognitive temporal maps for identifying the icons of comfort, belonging and commitment. We are now at the final stage, which analyses how all this new body of knowledge can be incorporated in planning practice. The intention here is not to analyse it on a technical 'one-to-one' basis but to discuss in principle how a knowledge different from the professional can be incorporated in planning practice and what the elements are that promote or prevent this type of incorporation.

The chapter begins with analysing planners' views regarding the validity of CBC in planning practice. It then takes one specific trend in planning, the 'old and new approach', and asks the extent to which this approach engages elements of CBC in planning. Finally it analyses case studies of planning in which notions of comfort, belonging and commitment are involved, looking at their relevance to urban planning and management.

CBC in planning practice – the planner's vision

The planner's vision regarding the relevance and contribution of CBC and other forms of local knowledge to planning practice is important because planners are the professional authority who lead the planning process. They have the power inherent in their professional knowledge to include or exclude other sources of knowledge in the plan-making process. We asked architects, planners and academics what their visions are regarding this matter:

> *'I am deeply suspicious of planners and architects who think that they know how to design the environment. There are so many individual buildings with symbolic cultural senses that deserve a great deal of attention, and for that planners and architects need special skills . . . I am sceptical about physical design without the social, cultural and political involvement of the users.'* (James, planner, London, 22 August 1999)

James is a planner and academic and he questions the ability of planners to incorporate in their work notions other then the physical. James suggests that planners and architects need to look at social, cultural and political circumstances in order to design the environment so that it contains and creates these elements. Similar opinions appear in the literature, suggesting that planners who do not have a broader perspective than the physical may cause the destruction and smashing of memory and belonging. Sandercock (1998a) is quite certain of the dominant role planners play in this process. 'Paradigms beyond modernist planning have essential parts', she argues, 'in acquiring and recognizing the importance of memory, desire and the spirit of the city in creating healthy human settlements' (p. 208). In her opinion the notion of belonging in the city is especially distracted, smashed and forgotten with modernist planning:

> Modernist planners became thieves of memory. Faustian in their eagerness to erase all traces of the past in the interest of forward momentum, of growth in the name of progress, their 'drive-by' windscreen surveys of neighbourhoods that they had already decided . . . to condemn to bulldozer . . . Modernist planners, embracing the ideology of development as progress, have killed whole communities and destroyed individual lives by not understanding the loss and grieving that go along with losing one's home and neighbourhood and friends and memories. (Sandercock 1998a: 208)

Indeed Sandercock's targeted criticism is the modernist planner who works with or in the service of the developers and produces plans that aim to meet developers' needs more than communities' needs. Kay Jordan, the Director of Spitalfields Small Business Association in London, has expressed similar opinions regarding the developers' intentions in the area:

> 'The community development plan for the Bishopsgate Goodsyard site was funded by the developers but working on a "community plan" was their way to avoid the heat that came up on the Spitalfields market site . . . this is not to say that the planner didn't do serious work but it was the community techniques that we all developed as architects and planners that were applied against us in an extremely divisive manner in order further not the community but the developers. How can you finish a community plan that is exactly the same plan as the developers'?' (4 August 2001)

The frustrations expressed by Jordan, herself an architect, derives from the much stronger economic power of the developers to finance the plan for the site they are interested in than that of the local communities to object or benefit from the plan. She is also protesting against the developers' manipulative acts of having a 'community plan' that is in fact a plan similar to what the developers want and expect to happen in the area and not really what the community needs and desires. The 'community plan', she argues, is a vehicle for the developers to reach cooperation with the community members and avoid the enormous objections submitted to the previous plan in the area against the developers' intentions of destroying the Spitalfields market and building offices instead, objections that halted the project for several years (see Chapter 4). However, there are different perspectives to any 'planning plot'. What does the planner of that community plan think? And what do the Bangladeshi people think?

The planner who prepared the Bishopsgate Goodsyard Community Plan obviously does not see himself as 'a thief of memory' or 'a killer of communities', on the contrary, he presents a more practical-oriented approach:

> 'What is fundamental to community plans is land and the physical development of land. Their context is about survival, my background is as a land surveyor and in economics and it is pointless to have a community plan if it is not succeeding. Land in inner London is worth millions of pounds and if I ignore it I do no service to the people because I'll be misleading them. Planners have got to be viable financially, politically, socially and economically . . . the context is survival. Communities are under threat of being displaced due to the workings of the property market. Any community plan does have to reflect their hopes and aspirations. After all they wouldn't want to stay there in the first place if they didn't feel they belonged . . . A community plan is a proposal that has to fight for its survival, so personal considerations [of comfort, belonging and commitment] help its credibility but should not compromise its viability . . . it's the balance that counts.' (Donald, London, 5 August 2001)

His approach takes capitalist development as part of an inevitable reality. It is the 'if you can't beat them you may as well join them' attitude he represents. In his view, concepts of comfort, belonging and commitment, although increasing the *credibility* of the plan may jeopardise its *viability* and thus they can not always be a part of a community plan-making process. The balance between the plan's credibility and viability is what counts and it seems that this planner perceives the plan's viability as more important then its credibility.

This practical approach brings us back to the discourse around the role of planners in the plan-making process that we developed in Chapter 2. Donald's view of his role as a planner is obviously not as a community mobiliser or transformer but rather as a mediator between the developer's interests and the community's needs. This seems to be a more practical approach of acknowledging the power of capitalism and thus taking the developer's interests as a base, as almost a reality. Donald reflects the matter-of-fact approach of what he perceives as gaining the best possible result for the community given inevitable spatial changes in the area resulting from developers' projects.

To complete the story of the Bishopsgate Goodsyard Community Plan it must be emphasised that one of the community leaders himself said that the Bangladeshi representatives as well had decided to choose the 'yes – but' option, that is, to negotiate with the developer over the community benefits (see Chapter 4). Some of the Bangladeshi members adopted the practical approach the planner advocates. They acknowledged the fact that changes would occur whether they liked it or not and therefore channelled their energies and resources to benefit the most they could from these processes.

These 'actors' represent three approaches to the situation. The first is the leftist approach that sometimes acts as a paternal protector 'not of the Bengali community *per se*, but of Spitalfields itself (Jacobs, 1996: 96). The second is the planner's approach, which is more practically oriented, accepting the existing capitalist power relations as a reality, and working out from there to benefit the community. The third approach is that of the Bengali people, which like any other community consists of a variety of opinions and interests but that officially adopted the 'yes – but' approach, accepting the developer's intentions but with certain conditions and thus objecting to the leftist approach of fighting against the developer's intentions at any rate.

Hillier (1998) discusses the planner's role in such complicated situations: 'A planner's representation of the Northeast Corridor tends to be Euclidean and instrumental', she argues when analysing the planners' role in a development process in the Swan Valley, Perth, Australia. This is an area that is claimed and used by Aboriginal people too. Using Giddens' perception she emphasises the role of planners as 'evacuators' of memory, claim and belonging.

> laypersons' knowledge embodies tradition and cultural values; it is local and de-centred. Planners' expertise, on the other hand, is disembodied, 'evacuating' the traditional content of local contexts, and based on impersonal principles that can be set out without regard to context – a coded knowledge that professionals are at pains to protect. (Giddens 1994: 85 in Hillier 1998)

No doubt the viability and validity of CBC in planning practice is a complicated issue. Most planners would agree to the definition of CBC as reflecting a good quality of life, a target that most plans aim to meet, but the challenge is how to incorporate this knowledge in planning practice. Let us first read the opinions and perceptions of some urban planners and architects who took part in the research:

'If you could find a planning system which could integrate CBC, it would improve social relations of the city life. It could be incorporated in the practice but planning is then not the physical but the social.' (John, academic urban planner, London, 11 August 1999)

'CBC are crucial in design. I use that in my work. To achieve belonging I would find what makes them [the clients] feel comfort, what kind of relationships they have. Comfort – I might dictate. Commitment – that follows.' (Janet, architect, London, 19 August 1999)

John and Janet both deal with planning. John is an academic who researched in the field of urban planning. He is also a practitioner who has been involved in several plans. Janet is an architect working mainly with designing houses and buildings. Their two approaches highlight the first important component in the discussion, *the scale* of the plan we are dealing with. John refers to the scale of city planning. Janet refers to the scale of house design. From the discussions on CBC in the six categories of space it becomes evident that the more 'private' the scale the easier it becomes to define and incorporate notions of CBC in the planning process, and the more 'public' the scale the more complex and challenging it is to define and incorporate notions of CBC in the planning process.

However, as mentioned in Chapter 1, the very basic thinking of the architect of the Israel National Master Plan, Shamay Assif, in formulating these concepts has been his own home. This is in fact the challenge that we face as planners in the age of globalisation: how to make our cities places where people can feel 'at home'. This means that planners have to work *beyond* the private/public dichotomies and scales and find out how to incorporate notions that reflect a sense of home in the city's spaces.

James raises another important point when discussing the relevance of CBC to planning:

'What is planning? Organised technical guidance for other people's decisions, but planners can't do it on the basis of power but knowledge. Planning has been criticised but it still exists. Planners have responsibilities and multiple objectives and sets of activities and planners are not much good at that. CBC in planning? It is a complex question. It depends in each society on the relationship between the economy and the free market and the relationships between state and civil society and the associations [that] exist.' (James, academic urban planner, London, 22 August 1999)

What is the definition and scope of the planning process is probably one of the crucial questions of this research. Urban planning in the context of this research is not only about the plan-making process. Rather, urban planning includes the implementation phase and the management of cityscapes as part and parcel of planning. From this broader perspective we can formulate new and different channels for incorporating CBC, not only in the plan-making process itself but also in the actual management of cityscapes. City planning, development and management is indeed a complex issue, as James himself mentions and it depends on the relationships between the free market, the state's controlled economy and the politics of planning and development. These elements affect the extent to which *civil society structures* are developed and *the roles that non-governmental organisations* play in disseminating and promoting such ideas.

Michal is an urban planner, an employee of one of the government ministries in Israel. It is interesting to read how she as a government planner views the incorporation of CBC in the planning practice:

> '*I think CBC should be taken into consideration. It must be taken into consideration in the plan-making process and this is my criticism about planning in Israel. CBC do have good solutions. Comfort is about accessibility . . . but it is difficult to provide it if you don't want to destroy existing buildings. Belonging and commitment are harder to achieve because it is difficult for a planner to design a city totally because it is a matter of its customers. For example, coffee shops, the planner can define the zoning but to open a café is a matter of demand. The authorities can provide incentives, the authorities can develop infrastructure and define land uses. That's how they contribute to creating CBC.*' (Michal, urban planner, practitioner, Jerusalem, 27 September 1999)

Michal highlights another crucial point in the discussion, '*Who designs and manages the city?*' What are the forces or the 'actors' that take part in shaping cityscapes? What Michal describes is a multiplicity of actors, which together contribute to cities' development. As a government planner she feels obliged to clarify this issues, and quite rightly from her perspective. She argues that the role of the official planning system in designing urban spaces is only partial. Like James she also mentions other 'actors', such as private entrepreneurs and the customers themselves, who are involved in shaping urban spaces. She also recognises the role of the authorities in city development, especially in providing infrastructure. As analysed in Chapter 4 the two powers – private entrepreneurs and authorities – do cooperate in the design of places and spaces in cities, sometimes blind to the needs and desires of the residents of those places and spaces. The weakening position of official apparatuses (state or municipality offices) that Michal presents in her narrative must be questioned. It seems that they are still the official mechanism used to promote citizen rights and needs. The municipality has the power to promote development by determining the level of tax rates that businesses pay. For example, the municipality can encourage the development of cafés and shops in the city centre by reducing their tax payments. This means that authorities still have the power to channel the development of the city in the direction they decide in spite of their declared weakening in the era of globalisation.

Ruth is a freelance urban planner. She works a lot with communities' involvement in the plan making process. For her CBC are obvious targets of planning:

> 'CBC are targets of planning. If a person doesn't feel she belongs and lacks comfort it is difficult to develop a good environment for her. Physical planning can incorporate CBC. For example in one of my projects a park was developed in the neighbourhood but the flats where the residents live do not face this park so the residents are afraid to use it. If the flats were facing the park it would be much more used and the residents would have felt much more comfortable and a sense of belonging. Another example is the concept of the cul-de-sac; it creates a feeling of unclear spatial division and thus a sense of discomfort. If people feel they belong they feel committed too. For example, graffiti is the expression of some sort of despair . . . of a lack of belonging and thus lack of commitment.' (Ruth, urban planner, practitioner, Jerusalem, 4 November 1999)

As a practitioner Ruth uses her experiences to clarify the viability and validity of CBC in planning. The examples she brings in her narrative are again a matter of the *scale of the plan*. She refers to neighbourhoods where she works with people. Her example of the park illustrates a neighbourhood-planning project, which has been drawn up without consulting the residents and caused misuse and negative perceptions even in technical matters such as the relationship between the flats and the recreation area. She actually highlights the importance of community participation in the planning process as a vehicle to incorporate CBC in planning.

Rebecca and Eitan's views represent yet again the connection between *the planner's politics* and the plan-making process:

> 'Obviously CBC are important in the plan-making process. It goes without saying.' (Rebecca, architect, Jerusalem, 3 February 2000)

> 'CBC are important as part of planning but different architects give them different meanings. I know some architects and all they are interested in is aesthetics – they make plans with no logic, skyscrapers that are there for their shape. They are aesthetic in the eyes of the planners, not the users.' (Eitan, architect, Jerusalem, 3 February 2000)

This connection between the planner's politics and the plan-making process brings us back to the very major point presented at the beginning on the section on *the role of the planner* in this process, that is, the context and content of the planners' views of their role in the plan-making process influence their attitudes towards the importance of CBC in the planning process. At the beginning of this section we presented Sandercock's criticism of modernist planners as 'thieves of memory', followed by Donald's narrative of his role as a community planner in the Bishopsgate Goodsyard project. His views of his role as a planner are completely different from the views of Rebecca and Eitan. Probably the latter could not function in the way Donald did in his negotiations with the Bangladeshi community in the Spitalfields area. This is an emphasis of the strong connection between the extent to which local knowledge is becoming part and parcel of the planning practice and the planner's politics and ideologies.

Last but not least, what does the co-planner of the National Master Plan of Israel, Shamay Assif, who had identified these definitions as components of a good quality of life, think of these elements as targets of planning:

> 'A planning process which defines targets and objectives makes the citizens feel more connected . . . You can create or increase a sense of belonging when there is an order in space, an order that one can understand, that this order is rational, a result of specific norms and attitudes. As long as there is chaos and developers build without any logic and the state's agencies put pressure no one can be attached. One can be attached to ideas: green boulevards, metropolitan areas, and major axes of movement, urban continuity. When you live in an urban tissue, in a plan, which is symmetrical, then you know where you're going. It's like your home that someone plans for you and you feel very attached to it and you do everything to realise it.' (29 February 2001)

The architect of the National Master Plan emphasises *spatial order* as an important goal of the plan-making process in increasing people's sense of belonging. This view represents an emphasis on the strong connection between CBC and city design and planning, which perhaps complements the other perspective elaborated in this book – of understanding the notions of CBC from their social or emotional sides. This emphasis on spatial order can be seen as the contribution of the professional planning knowledge rooted in the modernist view of planning. Assif also mentions negative forces and actors that create what he terms spatial 'chaos', which can be perceived as a result of the weakened power of the nation state, a situation described in the literature (see Chapter 2) as the prioritisation of efficiency over values, or of profit over welfare and private interests over public interests, a pattern that is becoming more and more significant in many cities in the world as a result of the processes of globalisation.

At this point I wish to pause and include in the text the comments of Shamay Assif on my own interpretations of his narrative presented above:

> 'To achieve CBC is a multi-faceted attempt that involves the identification of the rich mosaic of needs and aspirations of various groups and individuals through an effective participatory process. Many of these needs and aspirations often conflict with one another.
>
> It also involves devising integrative solutions and using multi-disciplinary tools to implement them.
>
> Solutions should combine social measures (that enhance, for example, community awareness and activity and empower local leadership and care for the

poor and handicapped), economic measures that create jobs, encourage business development and discourage polluting activities, etc., as well as physical measures (that enable easy access, enhance urban quality and develop friendly public spaces that work, etc.). All these measures should support one another to achieve an integrative planning strategy targeted to create more comfort, more belonging and more commitment.

But good planning should also be extremely legible and transparent and emphasise elements that are easily memorable, easily identified with, a real reflection of a collective perception of the environment one lives in and an authentic expression of [their] dreams and aspirations.

Diverse as they are these perceptions and expressions should be crystallised and integrated into simple policies and grasping images.

The new National Master Plan for Israel [Tama 35] supports the preservation of a National Green Axis, fosters four Metropolitan Regions and a set of lateral linear open space systems along Israel's major streams and *wadis*. The National Master Plan promotes urban life [strongly] and urban culture through quality compact urban development, rejuvenation of inner city neighbourhoods and business districts. It also puts high emphasis on the accelerated development of clean and fast mass transit systems that allow equal accessibility for all, calls for better pedestrian-oriented environments, as well as contributes to the stability and clarity of a simple memorable spatial order.

CBC starts with shared goals and objectives, it is enhanced when individual, social and cultural needs and aspirations are successfully met, and is celebrated when all these are expressed in a clear legible and memorable *spatial order*.

The chaos created by random pressures developed by competitive market forces or by conflicting public agencies is nothing to feel comfortable with, to belong to, let alone . . . of committing oneself to. This is true at the national level as well as at the urban level. It also relates with parallel notions to one's neighbourhood, the street one uses and one's own home.

A clear imaginable spatial order is something to identify with and hence obtain more CBC. We should clarify the notions of *chaos* and *order*. I perceive chaos as total disorder. This is when systems go wild and nothing is predictable. It is pathology of society and of cultural behaviour. Now order is a mere expression of culture. It should be seen as a framework for random activity, opportunity and excitement, but always a framework that supplies the security that is the base for all CBC factors. Order is when a mere *space* is made into *place* that expresses shared values, is identified with, and eventually contributes well to more CBC.' (22 December 2002)

These clarifications and elaborations on the points raised earlier help us to deepen our understanding of the complexities and the multi-layered nature of urban planning. The points are reflected in most opinions given above by architects and planners. But still what we can draw out of the analysis is that perhaps what is more significant in promoting ideas of civil society and democracy in urban planning is the actual influence of the politics of planning and development, which in many cases dictates the extent to which civil society can develop and quality of life of *all* people can be promoted in planning. Young calls it 'an ideal city life as a vision of social relations affirming group differences' (1990: 227). In Chapter 2 I elaborated on her ideas of 'being together as strangers', where she emphasises the necessity of a democratic life in order to maintain a 'healthy life' in the city. Democratic life means that 'city politics should not be dominated by the point of view of one group, it must be politics that take account of and provide voices for the different groups that dwell together in the city without forming a community' (p. 227). Here again the concept of 'citizenship in the global city' or 'multi-layered' citizenship becomes relevant, acknowledging people's citizen rights of comfort, belonging and commitment as part and parcel of civil society. The right to comfort and the right to belong contain in their definition the right to difference. A sense of comfort and belonging for people of different identities means that their differences are acknowledged and respected.

To conclude this section, from the planner's narratives above some major conclusions and suggestions can be drawn as to the viability and validity of CBC in urban planning, development and management:

- *Credibility and viability and the role and politics of the planner.* It seems that the role of the planner in the age of globalisation is not at all easy. They have to manoeuvre between designing a plan that will be credible, that is, reflects residents' needs and especially desires. Sandercock (1998a) makes a clear distinction between *needs*, which can be identified in 'needs surveys' and *desires*, which involve 'the subconscious, a personal engagement, dreams and feelings, an ability to intuit the atmosphere and feeling of a place' (p. 212).

  At the same time the plan has to be viable, that is, meet the interests of the decision makers, state planners and developers. This is a complicated situation for planners and is perhaps the reality for most of them in the current era of global economic forces, which puts efficiency at the forefront. Credibility and viability is one way to deal with such situations, as Donald faced in the Bishopsgate Goodsyard project. It may become more common for planners to deal with such situations in the cities of tomorrow and therefore it is worth understanding such circumstances and perhaps clarifies the planners' role in such negotiations between communities and the authorities or the developers. As mentioned in the text above and previously in Chapter 2, planners who find themselves in such situations may face a dilemma as to their role and their politics in 'the planning game' that builds up between private entrepreneurs and communities. The extent to which planners are willing to jeopardise principles of CBC or putting in questions on people's needs and desires depends on the ways planners develop their professional identity – either as mobilisers or social transformers or strictly as technicians.

- *The scale of the plan.* Obviously the scale of the plan influences the extent to which the CBC can become targets of planning. The smaller the scale of planning the greater the chances and possibilities of incorporating the CBC of individuals in plans. As we have seen previously the 'home' scale is the less complicated to identify when deconstructing notions of CBC. This is the scale from which the idea of incorporating these elements as quality of life were drawn (see Chapter 1). It is at this point that it becomes a challenge to transform the characteristics of CBC, which are so obvious at 'the home' scale to 'the city' scale. At the end of the day the challenge that we face as planners in the age of globalisation is precisely how to make our cities a place where people can feel 'at home'. The case of belonging and spatial planning in Israel clarifies the significance of the 'scale' in incorporating these issues. The first step towards this direction is the acknowledgment of decision makers and planners of the importance of concepts such as CBC in improving the standard of living in urban spaces. Some more practical suggestions appear later in this section.

- *What is the definition and the scope of the planning process?* As already mentioned, urban planning incorporates much more than the plan-making process itself. It includes the implementation phase and the management of the city as part and parcel of planning. From this broader perspective we can perhaps formulate new and different channels for incorporating elements such as CBC (or others) not only in the plan-making process but also in the actual management of the city streets, neighbourhoods and centres so that they include components of belonging and attachment and thus these spaces develop into greater places of comfort and of commitment.

We need to realise that a deeper involvement of local knowledge is needed both in the actual planning process and in the management of the city. A greater level of involvement means not only using the well-known and heavily discussed techniques of participation in the various planning stages. It involves a deep understanding by planners and decision makers of the importance of incorporating notions such as comfort, belonging and commitment in the planning process so that the whole procedure of planning would be targeted to promote such notions as quality of life. In Chapter 2 I suggested that even the methodology of the rational comprehensive model can become a tool for incorporating local knowledge in planning. Each of the planning stages (data collection, formulation of planning goals, designs of alternatives, choosing the preferred alternative, designing the alternative, etc.) can become a space for negotiation and a mutual learning process between communities and planners on the various elements of local knowledge.

The level to which communities and individuals can raise their voices in their efforts to promote elements of quality of life depends on the extent to which civil society's mechanisms and norms allow such interventions and involvement. The role of NGOs becomes crucial in such negotiations. These organisations can play an active role not only in advocating norms and values of quality of life in the planning process but also and perhaps in particular in the daily and routine management of urban spaces. As will be shown in the next section, NGOs and grass roots organisations are becoming the 'watchdogs' of guarding and promoting civil rights in the field of urban planning and city management.

One of the examples of creating spaces of belonging and a sense of place within a wider scope of planning is the planning and management of Banglatown in London. This area consists of spatial elements of belonging and sense of place for the Bangladeshi people who live there but at the same time these spaces are threatened by developers. These activities and events are part of the whole apparatus of urban management in which urban planning is only one part.

- *Civil society mechanisms and the role of NGOs.* So far the discussion on planning, development and management has focused on the state and its roles and power in incorporating notions such as CBC in planning practice. However, the concept and structure of civil society suggests alternative ways of action initiated by communities and associations that could promote and support ideas such as the incorporation of elements of quality of life (CBC or others) in planning, development and management of cities. The role of NGOs in such processes is significant. As seen later in the case study analyses, in both the discussion on belonging and spatial planning in Israel and in Banglatown, the role of NGOs is crucial in promoting the interests of the various communities involved, interests that they define as part of their quality of life. NGOs' independent character allows them to promote communities' needs and desires that the state and sometimes municipalities tend to neglect.

  As we saw people in both cities associate a sense of commitment with such activities. Both in Jerusalem and in London people connected commitment with activism and involvement in NGOs' fights. People's sense of awareness towards their own needs and rights is an important part of developing civil society's mechanisms. And thus the role and position of NGOs becomes central and crucial for incorporating and maintaining norms and values concerning quality of life in urban planning and management. Their increasing role and power can be perceived as a sign of the weakening of the nation–state in that their activities have become an alternative channel of promoting and strengthening principles that are significant and even crucial for individuals and communities, be it principles of CBC or other terminologies of quality of life.

- *Who designs the city in the age of globalisation?* The new priorities of planning in the age of globalisation have been identified as the speeding up of the turnaround of planning applications, the completion of local plan preparation and the streamlining of procedures such as public consultation, which potentially slow down the plan-making process (see Chapter 2). The end result is that cities in this era are designed and managed by many more 'actors' and interests than before, not all of them necessarily seeing the importance of promoting and incorporating principles of quality of life such as comfort, belonging and commitment in the planning and management of cities. This situation has become more explicit in certain cities such as Jerusalem, where the interests and needs of certain sectors in the city are met at a much greater level than other sectors. In Chapter 4 we discussed the discrimination of Palestinian residents of Jerusalem as one such an example. In London, the dominance of the developers in reshaping cityscapes has increased in the last two decades and economic interests are sometimes prioritised over communities' quality of life.

In such situations of multiple 'actors' and 'interests' there is a greater need to determine priorities and 'red lines' so that cities will be home for their communities and residents. To maintain such a sense of home in the city the mechanisms of civil society should be dominant in the planning and management of our cities. This means first and foremost that the power of the authorities to determine planning rules and principles and to take the leadership in designing the cities of tomorrow should become dominant again in spite of the weakening of the nation–state and other centralistic apparatus in the era of globalisation. City's bureaucrats should use their power by law (by issuing planning permits) to make sure that cities become more comfortable, maintaining people's sense of belonging and making their citizens feel committed to their city. NGOs and other informal organisations should play the role of guards so that policies and plans meet the needs of the various communities living in the city. This balance between cities' growing and expansion and their function as 'home' to their residents should be maintained by both the city bureaucrats and the NGOs as the protectors of such balance.

- *The practicalities of incorporating CBC in planning, development and management of cities.* The practical suggestions as to how to guard the quality of life of residents in the cities of tomorrow becomes an interaction between three main domains: that of the practitioner, that of the academic and that of the activist.

The first arena concerns the practicalities of urban planning and management. It is about the incorporation of what I term here a *'Community Impact Assessment'* to become part of the requirements needed for developers to attain building permits. This Community Impact Assessment will be carried out by independent planners and community mobilisers who will work with community members on the predicted impacts of the project on people's own definitions of what is for them quality of life.

In this book we have adopted the terminologies of comfort, belonging and commitment as part of quality of life but this could be only one set of norms and values of quality of life. Communities of different cultures may have their own sets of values and norms for defining quality of life.

A Community Impact Assessment becomes then a practical tool to evaluate the cost and benefits of developers' and municipalities' projects for the community members from their own local perspectives. It is perhaps a more precise and a more 'grassroot' framework than the Social Impact Assessment in that it looks not only at general social indicators but also at specific measurements that communities themselves have identified as consisting of their quality of life. The Community Impact Assessment is based on the local knowledge of people in evaluating the extent to which proposed projects could become obstacles to people's daily lives.

If such assessments could become part of the requirements needed to obtain planning permits these procedures could become an alternative to what developers do now, when they pay planners to become mediators between them and the communities, which might turn out to be a piecemeal persuasion and cooptation process accompanied by promises from the developers to invest money in various community facilities – promises that are not always kept (see Chapter 4 for the Spitalfields case). Community Impact Assessment will be initiated prior to the provision of the building or planning permit and will include legal obligations on the developers to make sure that the quality of life of communities is not jeopardised as a result of the suggested project. The same goes for projects initiated by the authorities; part of the requirements needed for the project to be approved would be the Community Impact Assessment, which will evaluate the effects of the project on people's lives.

The second suggestion to promote notions and ideas of quality of life in planning, development and management is related to academic research. Academics who work in the field of planning and development play their part in developing new ways of understanding and methodologies of incorporating the diversified notions of quality of life in planning, development and management programmes. Such work is currently ongoing among several academics in the field (Leonie Sandercock, John Forester, Patsy Healey, Oren Yiftachel, Jean Hillier to name only a few) – all working on new understandings and new ways of thinking with the aim of exploring the complexities and basic discourses that are the grounds of urban planning and development in the twenty-first century.

The third way to guard notions of quality of life in urban planning and management is by exploiting the options existing in civil society's forms in order to influence and affect our daily spaces. The role of NGOs and other grass roots organisations in such channels has already been discussed. It is worth mentioning again that in fact the power and stability of civil society depends on the cooperation between these three domains: the practical, the academic and the activist.

After pointing out some of the major issues concerning CBC in planning practice it is now appropriate to discuss the connection between CBC and planning practice in detail. First comes the discussion of the extent to which the 'old and new' approach indeed incorporates notions of CBC. This is followed by a detailed discussion on the relevance of these notions in three case studies.

'Old and new' in urban planning and management

'Jerusalem makes me feel comfort and as if I belong. There is something about its history that makes me feel connected to this space. This is why I've chosen to live in Jerusalem although I work somewhere else . . . Buildings preservation is very important. History shouldn't be destroyed, something that "they" do a lot in Israel. City preservation makes me feel I belong, when I see that "they" invest [money] in order to improve [the city] it makes me feel good, a feeling of freedom and movement, when I see Jews and Arabs using the same places it makes me feel good and as if I belong [having] communal activities, a feeling of political "maturity" makes me feel good and as if I belong and pride – which is very rare. A spatial respect for the "Other"' (Mary, 40s, single, Israeli – Ashkenazi–Jew, Jerusalem, 4 August 1999)

Mary's local embodied knowledge of her daily life in Jerusalem links between notions of comfort, belonging and commitment and the politics of urban planning, development and management. Mary was interviewed in 1999 when there was still hope for a better future for the city. She mentions the history of the city as part of her sense of belonging. She connects the history of the city to urban planning and management by talking about building preservations – the city of memory in Sandercock's (1998a) terminology. Like many others (as seen in Chapter 6) Mary also mentions the importance of preserving history in creating a sense of belonging. She also refers to the major role of the authorities – the government and the municipality – in formulating plans that will respect the old buildings of both Jews and Palestinians and so increase the sense of belonging of the city's citizens. She also mentions what was termed in Chapter 3 the 'spaces of citizenship', that is, public spaces which respect equal development and equal investment in both West and East Jerusalem. This is in her view what brings mutual use of space, perhaps it relates to what Innes (1995) terms 'a matured democracy'. Unfortunately at the time of writing the situation in Jerusalem is far from Mary's 1999 vision and it is far from becoming 'a matured democracy'.

The old and new approach has gained more awareness among urban planners and architects in the last decade or so, especially with regard to public buildings. It might therefore seem somewhat banal to discuss the old and new approach in connection with promoting a sense of belonging and attachment to the city. Nevertheless it is important to emphasise the different faces of this approach. While it advocates the preservation of the 'old', which is 'public', it ignores and tends to neglect the 'old', which has no historical 'public' significance, but does have significance for certain communities living in the city.

The old and new approach has several expressions. First, its target is to incorporate new architectural forms in old buildings that are significant in the history of the city but no longer meet the growing needs and demands of visitors to these sites. As already mentioned in Chapter 4, several examples exist in London: the Shackler Galleries at the Royal Academy of Art and the

British Museum Great Court, both designed by Sir Norman Foster. Sir Norman himself related to this approach in an exhibition titled 'Exploring the City – The Foster Studio' which took place in the summer of 2001 at the British Museum:

> The combination of old and new buildings in cities creates a rich and inspiring architectural environment. Historical buildings add character to our cities, but many of them no longer suit the purpose for which they were originally designed . . . however, rather then being declared redundant they can frequently be given new life by sensitive modern additions. Larger, complex old buildings are like the city in microcosm, the products of generations of growth and change. Just as our cities benefit from the dynamic mixture of historical and modern architecture, so too the combination of old and new within a single building can add richness. In the process, we can uncover and reveal layers of history that add to our enjoyment of these structures. (Exploring the City – The Foster Studio, British Museum, 22 June–7 October, 2001)

Another expression of the 'old and new' approach is making a new and different use of old buildings that no longer serve their original purpose. Rather then demolishing them they are reused for new purposes. The most popular example is the Tate Gallery of Modern Art designed by Herzog and de Meuron in the Bankside Power Station, situated just opposite St Paul's Cathedral on the Thames in London (see Figure 10.1). Bankside was built during the 1960s and served as a power station for several decades. This redesign of the power station as a museum takes advantage of the existing space and its enormous turbine hall. There are other examples of new construction, especially in the City of London, which preserve some or parts of the previous, old buildings (see Figure 10.2).

In Jerusalem, this 'old and new' approach has been implemented in particular in the construction and perhaps reconstruction of public buildings such as the new city hall compound (see Figure 10.3). While the city hall building itself is new most of the surrounding buildings date back to the Ottoman and the British Mandate periods (except the French consulate building, which has been demolished). The old buildings were refurbished and serve both as part of the municipality offices and as business and leisure areas with restaurants and cafés. The connection with the past is made with a Blue Plaque attached to each of them (see Figure 10.4). This 'old and new' approach is one way to maintain 'the city of memory', to use Sandercock's terminology, not only private memories but also or perhaps mostly communal memories that link space with past events and create a sense of a linked existence between the past history and those who live in the city today.

However, as mentioned above the 'old and new' approach is usually used with regard to public buildings or those with public interest or a significant historical past. But what about the preservation of buildings of less historical or national importance to the collective memory? These buildings sometimes have a unique character – usually architectural or local – and they are essential in maintaining the spirit of neighbourhoods, streets and a sense of place for those who live their everyday lives there. Here the real and perhaps harsh tension

Figure 10.1   The view through the '**d**' of the Tate Modern art gallery, London, towards St Paul's Cathedral

between municipalities' and developers' intentions on the one hand and residents' wishes on the other have their most explicit expression. Some of the biggest fights against developers' intentions can be labelled as those between 'capitalism and comfort, belonging and commitment', as cynical as it may sound. One such example was presented in Chapter 7, when Ronit illustrates what is a sense of commitment for her by talking about the fight against the developers' intentions of destroying the natural park near her neighbourhood in Jerusalem and building skyscrapers there. For the residents of the neighbourhood the park is a source of comfort, it increases their sense of belonging to the area they live in, but what are labelled 'public' interests are used as a token in the planning game and the capitalist intentions of the developers sometimes win over the quality of life of the local residents. In the case of this park the battle at the time of writing was going in favour of the residents – the

Figure 10.2 'Old and new' in the construction process, City of London

regional planning committee accepted the residents' and several NGOs' objections to the plan – but unfortunately this is not the end of the story, as the developers themselves promise.

Another example is the battle of many organisations in the United Kingdom and Israel for preserving not only historical buildings but also those that have a significant architectural style and importance to their close environs. In Spitalfields, it is the famous battle of the Spitalfields Historic Trust to save some 230 Georgian houses. This can also be perceived as a battle against demolishing sites of belonging and of memory, not so much with historical reference but mostly with architectural importance. In this battle between communities' needs and desires for preserving the 'old', for maintaining their sense of belonging, and the developer's capitalist ambitions, the power of the latter has proved stronger. The recent developments in the Spitalfields–Banglatown area are one

Figure 10.3 'Old and new' in the municipality's compound in West Jerusalem

Figure 10.4 The Blue Plaque with historical details, municipality compound, West Jerusalem:
    Bergheim House, built in the 1880s as the home of banker M.P. Bergheim. In the 1920s it functioned as the Hotel De France, later as the Detroit Hotel. From 1950 up to the 1980s it served as the laboratory of the Hebrew University. It is now a restaurant.

such example. As already seen in Chapter 4 this area has become a high priority for the developers because of its proximity to 'the City' and the growing demand for office space and homes. The public–private partnership system in the United Kingdom through which the government and the municipality share common interests with developers to maximise the potential return from sites in the city decreases the chances of putting people first. The Spitalfields Market project, which attracted tremendous objections precisely because of the need to retain a sense of belonging and attachment by preserving the old market, is now in its first stage of construction. It is quite evident that part of the market is destroyed in spite of big advertisements announcing it as saved (see Figure 4.5). It seems that communities themselves have to fight for realising their own quality of life and in most cases with not too much success.

Let us now present four cases in which communities' commitments to notions of comfort, belonging and commitment will be emphasised; the first case represents a universal experience of discomfort that exists for women in most public facilities around the world. The other three cases represent different locations from Israel, London and Cape Town, South Africa.

On CBC in planning practice – case studies

This last section questions by way of example the relevance of incorporating local embodied knowledge of people in creating and maintaining spaces of comfort, belonging and commitment in globalised urban spaces. Some cases, such as the first, one are an example of a universal lack of awareness and the blindness of planners to the unique needs of people, which has caused discomfort. The other cases are more complicated and necessitate a different way of thinking about the politics of planning and development in order to be able to make a change.

On lack of comfort in women's public toilets

While writing one of the book's drafts in London in 2001 and being preoccupied with understanding the meanings of CBC and their relevance to city planning, I heard the following comment made by a woman waiting in front of me in a long queue for the women's toilets at the Royal Albert Hall during the Proms Season. She was angry and I imagine she also felt discomfort: 'This [the toilet] wasn't designed to cope with people.' I thought to myself that this woman expressed very clearly the lack of awareness and the blindness of architects and planners to the basic human needs of women in their most straightforward meanings. By their lack of fulfilment a great deal of physical discomfort is created.

I believe that all women who have used public spaces have experienced the unpleasant, everyday practice of queuing for women's toilets while men's toilets are usually empty. Much has been said about the biological differences between women and men that make urination a much longer process for women then for men. Women need personal cubicles to urinate; men can do it in the public space of the toilet. This means that the act of urinating takes longer for women than for men and therefore women's toilets are busier then men's toilets. This basic biological difference obviously necessitates a different approach to design, maybe through allocating more space for women's toilets than for men's so that more women's cubicles can be constructed. But somehow these simple and basic facts do not reach the minds of the architects and they carry on designing the same size toilets for men and women and thus create physical discomfort for women. Women sometimes protest. I have seen queues so long that the women go and use the men's toilets. This simple but very clear example emphasises how a consideration of gender biological differences could solve an enormous sense of discomfort for women in public spaces such as theatres, cinemas, cafés, concert halls and parks.

CASE STUDY   On belonging and spatial planning in Israel

On 12 August 1924 at 11.30 am a British aeroplane took off and from an altitude of 6000 ft one of the crew took several aerial photographs of Palestine, then under British Mandate. An enlargement of sheet no. 2 of that aerial photograph, entitled Jaffa, now hangs on the Western wall of the Department of Geography, Tel Aviv University. Students and staff members can clearly observe the agricultural area that existed East of Jaffa with particular reference made on the sheet (in English!) to the locations of three Palestinian *Saknet* (Arabic for neighbourhoods): Saknet el Araine, Saknet Abu Kabir, Saknet Hammad. The same area looks totally different today; instead of the Saknet the area is populated with *Schunot* (Hebrew for neighbourhoods), Jewish neighbourhoods with different names and different memories. The sharp contrast between these two images emphasises more than anything else the complex situation of Israeli society and space today. The state of Israel was established 55 years ago as a sanctuary place for the Jews, 'the gathering of the exiles', from all over the world, especially for the survivors of the Holocaust from Eastern Europe and the victims of anti-Semitism encounters in Arab countries. But this 'Promised Land' was not empty; some 700,000 Palestinians lived in Palestine before 1948, some of them for many generations. Still, after 55 years of independence, it seems that the state of Israel does not feel strong enough to deal with these sensitive historical facts and expresses reluctance to allow any symbolic or spatial expressions of belonging, attachment and memory to the Palestinian–Arab citizens of Israel.

Is a sense of belonging to a space an either/or situation? Either for Jews or for Arabs? Is it possible to construct belonging to more than one nation or one group of people? Can spatial planning create or relate to a sense of belonging? And if yes, how can we commemorate a sense of belonging using planning? These are some of the questions that this section aims to challenge. In order to clarify and perhaps simplify this complex situation I have chosen to focus on a very localised planning plot, which illustrates the complexities of discussions on the notions of belonging, memory and attachment in Israel. It is the case of the Tomb of Az-A-Dean-El-Kasam.

The tomb of Az-A-Dean-El-Kasam or the case of the road and the graveyard

Nations are imagined (Anderson 1983) but even these imaginations are in many cases spatial. Nationality in Israel has strong spatial expressions and it is very much territorial. The sense of nationalism expresses itself in the formation of belonging and attachment to a place based on memories, nostalgia and claim.

One of the Palestinian villages that was abandoned after 1948 was Balad-Al-Sheik, situated near what is now the Jewish Local Council of Nesher. The village does not exist any more but its graveyard does, and it became quite famous in the 1990s because of the tomb of the Sheik Az-A-Dean-El-Kasam. This sheik was originally a Syrian who fought against the French occupation in Syria. He escaped to Palestine in 1921 and started his combat against the British because he claimed they helped the Zionist movement in its struggle against the Palestinians. He himself was not involved in any conflict against the Jews, only against the British, and he was killed in 1935 by the British and was buried in the Balad-A-Sheik graveyard.

In the 1990s the Islamic Hamas office in charge of the guerrilla fights against Israel adopted his name. His tomb became a contested space. For the Palestinians, both citizens of Israel and those living in the West bank, it became a place of pilgrimage, with the Hamas movement taking care of cleaning it and surrounding the tomb with a big fence (see Figure 10.5). For the Jews it became a disputed place associated with the Palestinian killings of Israeli Jews by the terrorist group Az-A-Dean-El-Kassam. No wonder that in such a bitter and complicated atmosphere the plan of the local council of Nesher to expand the road and expropriate some of the graveyard's area caused strong objections and demonstrations among the Arab religious movements in Israel. The conflict became more heated when the life of the Jewish head of the local council was threatened, mainly because of the declarations on the Jewish side that the sheik was not buried in the tomb. A committee was set up to negotiate between the two sides, chaired by the head of the centre of local authorities in Israel. Members of the committee were Muslim public figures, government representatives and the Nesher local council representatives. A compromise was suggested: to move the road so that only one tomb at the corner of the graveyard would be removed. The Muslims rejected this proposition and applied to the District Court. The District Court rejected the Muslims' appeal (Berkovitz 2000). In the end the road was constructed on columns, a compromise that satisfied both sides (see Figure 10.6). This event raises many questions as to the role planning and development play in commemorating and honouring sites of memory and belonging, in this case the memory and sense of attachment of the Palestinian citizens of Israel as related to a figure with contested associations.

Clearly these events show a strong Jewish claim to the area versus a strong objection to respecting or considering the notions of belonging of the Palestinians. But the end of the story shows that perhaps it is time to develop concepts of mutual belonging. This drive to commemorate memory in order to emphasise belonging is what motivates many Arab citizens of Israel to celebrate

Figure 10.5 The Tomb of Az-A-Dean-El-Kasam, Balad-Al-Sheik graveyard, Nesher, Israel

Figure 10.6 The compromise solution – the road on the bridge and the graveyards underneath

the day of 'El Nakba' on Israel's Day of Independence and to conduct memorial ceremonies in some of the villages abandoned in 1948. Being excluded from the hegemonic national Jewish identity, which in itself built on symbols of memories of biblical times, the Palestinians construct their own identity and sense of belonging based on their own memories. The developments in this planning story show that both sides use legislative means to achieve their goals. Both the Jewish and the Palestinians used the court and both sides were backed by different organisations, some of them NGOs. It seems that only when all other options were exhausted could a compromise be reached. The compromise that has been achieved seems like a result of goodwill and maturity as to the future of the mutual Jewish–Arab coexistence in the area.

Belonging and a sense of place in planning practice

To what extent can planning play a role in maintaining and respecting spaces of memory, of desire and of spirit – spaces of belonging? Whose belonging do we commemorate?

The case of the tomb of Az-A-Dean-El-Kasam is a local one, a dispute that has been going on for several years and has been resolved in what seems a mature compromise. But what about commemorating or smashing memories and the sense of belonging on a national scale of planning? I posed these questions to the architect of the Israel Master Plan (*Tama/35*) – Shamay Assif:

> 'There is a need to create a situation in which the Arab population feels they belong with this country, maybe there is a need to tell the Arabs that the state respects and is going to restore some of their abandoned villages and make them into memorial sites but we did not suggested it in the National Master Plan . . . We live in a complicated situation of a constant threat, it is a matter of survival . . . the two nations and the two communities should become more mature and more integrated, with a wide agreement on what quality of life means, but to suggest to restore their villages and the second stage is (throwing us to) the sea? . . . It is very popular to say that we are strong enough and we can allow ourselves [to let the Palestinians restore their belongings] but we are not so strong and we can not allow ourselves . . . there is a need to respect and to take care of the memory of the others but between that and solutions of [restoring villages] there is a big gap. For example, the tomb of Az-A-Dean-El-Kasam, who is a Palestinian hero . . . there are situations in which a proposal to commemorate his tomb will be rejected so I told my team not to include this site [as a site of historical preservation] but we did suggest restoring other sites in the National Master Plan, such as Kafar Kana or Karnei Hittin, where Salah-A-Dean fought against the crusaders, and this in itself is a revolution . . . to insert such sites into the National Master Plan. I told my team that if we included the tomb of Az-A-Dean-El Kasam as one of the sites for preservation the plan wouldn't be approved – we have to be practical. There is a plan – there is politics and there is reality.'
> (March 2001)

It is this same architect who defined the notion of comfort, belonging and commitment as formulating quality of life in the Master Plan who faces this dilemma. His dilemma concerns preparing a plan that expresses the hegemonic consensus of the Jewish, a plan that will be acceptable by the authorities, what we have termed before a *credible* plan. But the architect also wishes to formulate a plan that will be as pluralistic as possible, that is, a *viable* plan. This is actually the same dilemma that Donald, the community planner of the Bishopsgate Goodsyard, raised earlier in the chapter: a dilemma between values and their effect on the credibility of the plan. At the end of the day it is a Jewish planner who decided not to include the site of the tomb of Az-A-Dean-El-Kassam as a site for preservation in the Master Plan because he thought the plan would not be approved had he included it.

Here I pause again to add the clarifications and comments of Shamay Assif to the above analysis:

> 'The National Master Plan introduced the notion of belonging and memory through the designation for preservation of certain urban and rural fabrics and ensembles. Each group in the Israeli society (religious Jews, non-religious Jews, Christians, Druze, etc.) obviously has a different set of preferences as to which one's fabrics and ensembles should be preserved.
>
> There was the challenge of the plan that had by its basic definition to express a high level of consensus and on the other hand express different sets of values and priorities.
>
> We applied two decisions to bridge these conflicting requirements: one promoted the pluralistic approach which raises the understanding that respecting the needs and aspirations of others is a common value by itself and should clearly be expressed.
>
> The other is to avoid designation of individual, very specific sites (as opposed to larger fabrics and ensembles), especially those that represent and symbolise actual threats by one group or another. Such sites are well outside the consensus and hence perceived as harmful to the whole process.
>
> The nice thing about the plan is that it managed to put hundreds of sites within the pluralistic understanding, sites respected by different groups, but nevertheless within the consensus.
>
> Yes it excluded sites such as the grave of Az-A-Dean-El-Kasam, whose name is used as one of the most aggressive and active Palestinian terrorist groups directly responsible for the cruel murders of innocent civilians. Actually no one seriously suggested its inclusion. Even Arab members of the team and steering committee well understood its negative effect and limited value.
>
> This is like including the grave of the Jewish extremist Baruch Golgstein who cruelly murdered a whole group of Muslim Palestinians while praying in Hebron. There is a group of extremists that would even to this day support such an idea.
>
> It is not just a matter of mere viability [the ability to get the plan approved by the authorities]. It is a matter of whether the plan really contributes to the CBC or just promotes more conflict and less security.' (22 December 2002)

This discourse and the sensitivities regarding such disputed issues that are reflected in Assif's comments in his previous narrative and on my analysis express both the internal dilemma of the planner in such complicated situations and highlight the general issue of the role of the planner in establishing the 'boundaries of difference'. Identifying 'the boundaries of difference' within the context of the hegemonic Jewish belonging and memory can be a very complicated matter. It relates to the extent to which the planner functions only as a 'technician' who merely implements policies and the extent to which they have the liberty and freedom to influence and change.

Let us analyse this case by way of conclusion by relating to the points made earlier in the chapter, looking at the viability of CBC in planning practice. This case clearly illustrates how politics of planning and development dictate whose expressions of memory and belonging are commemorated. But beyond that it shows how significant and also complicated is the role of the planner in such a situation. The planner of the National Master Plan is the 'knower', especially for the authorities, he is the professional expert. At the same time he is aware of the fine line between professionalism and politics in planning, especially on the *national* scale. The planner is aware that the practicalities of incorporating sites of belonging and memory of one group in the plan actually diminishes its credibility. Like the planner of the community plan in Bishopsgate, the architect of the master plan poses the same dilemma between the *credibility and viability* of the plan. Here the question of *who designs spaces* becomes evident. It is not only the planner's field but also the authorities, and other interest groups – both Jewish and Arab – that are involved in such a planning process. This involvement of local groups and NGOs highlights another point that is important in the discussion of CBC and planning – that of *civil society mechanisms and the role of NGOs* in promoting various citizenship issues. Here local organisations obviously play a major role in such processes and perhaps even in the final decision of the local council not to destroy any graves but to build the road as a bridge. This case shows how *scale* matters in incorporating concepts such as CBC in planning. It seems that at the local level it is sometimes easier to bridge the gaps between two sets of contradictory memories and belonging. The practicalities of everyday life sometimes push for a compromise, which is more difficult at the national level. Here the hegemonic ideology plays a major role in determining whose memory and belonging are expressed in planning documents and the planning process is targeted to achieve a consensus so that the authorities can approve the plan.

This case presents dilemmas more than it suggests solutions. As planning is usually very political, especially in disputed places such as Israel, the articulation of notions of belonging and memory of the 'Other' are becoming very problematic, and the extent to which these needs and ideologies can become part and parcel of the hegemonic planning process depends on the degree of maturity and confidence that the 'majority' hegemonic society has itself to allow memory and belonging of the 'Other' to be part of the planning process and the civic agenda of the society as a whole.

CASE STUDY   On belonging, city planning and management in Banglatown

The Brick Lane area is an example of how 'spaces of belonging and memory' are established as part of a broader development scheme. The neighbourhood can be seen as incorporating notions of spatial memory and belonging but there might be other interpretations of these elements, as explained below.

Turan is a Bengali man who lives and works in the Brick Lane area in London. He says he sometimes feels that Brick Lane is part of Bangladesh. The reason he and other interviewees feel this is because it is a neighbourhood where most of the residents are Bengali, where most people came from the Sylhet district in Bangladesh. It is a community with strong spatial expressions of social networks and ties. Supermarkets contain local food, shops have traditional Bengali and Indian styles and music shops with their loudspeakers create a unique atmosphere. Recently, the names of the streets were put in the Bengali language together with other street signs and street lights were designed in the Hindu style (such as the symbolic 'gate' to the neighbourhood at the entrance to Brick Lane Road (see Figure 6.2).

These 'ethnic spaces' began to function out of local groups' demands for basic food ingredients, clothes and music that would meet their ethnic and cultural needs. There are many ethnic areas such as this in many cities in the world, including the very famous examples of 'Chinatowns'. In the 1980s these 'symbols of belonging' were targeted for commercialisation in the Brick Lane area. The design and transformation of these 'spaces of belonging', which were primarily spontaneous, into a commercial 'Banglatown' were part of planning – part of the Spitalfields Community Development Group plan (1989). This was a group of local Bengali established by the male business sector in order to influence but not halt the proposed development in Spitalfields. Their plan suggested developing the multicultural nature of the area with the focus on commercial and cultural life, a bazaar area and shops, restaurants, showrooms and craft shops. This commercialisation would represent the diversity of cultures in the area given the range of Bengali, Jewish, English, Somali and other ethnic groups living there. The plan would attract tourists to the area but would also serve as a reinvention of 'back home' (Jacobs 1996). This use of 'ethnic spaces' in cities especially for food and fun has been identified as serving to ensure cultural hegemony in multicultural societies (Anderson 1991). Places such as Chinatowns were created at times of negative stereotyping of both the Chinese and Chinatowns and served as the representations of the Other. Lately, with the development of nation–states' commitment to multiculturalism the idea has become more positive (Anderson 1991).

The idea of Banglatown was not seen as a simple return to the ancestral past and sense of belonging. Nor was it an appropriation of external capitalist interests. It was, as Jacobs argues, 'an activation of an essentialized identity category by one sector of the Bengali community – the businessmen, within the terms of the enterprise-link development opportunities available' (p. 100). Banglatown was an idea that helped to sell the economic opportunities of the area. 'Banglatown formed the cultural framework around which alliances could be made between big business and Bengali small business' (Jacobs 1996: 100).

But does this make the area less authentic or with a lesser sense of belonging and attachment? Obviously every 'ethno-town' begins in response to the demand of the local communities for ethnic and cultural consumerism. Grocery shops are opened with special food and spices that people of a specific ethnic background demand. Food and spices are some of the ingredients that make people feel comfort and as if they belong, as we have been from people's narratives.

The concept of Chinatown and Banglatown is developed only in areas that have the base for producing spaces constructed by ethnic identities. For the purpose of discussion it does not matter whether the latest developments of turning

Spitalfields into Banglatown is a matter of planned intervention or not, what matters is that the foundations were put there by the people themselves, the Bengali at present and the Jews and the Irish in the past. And that the 'controlling planning powers' did not prevent them from developing this sense of belonging in their area. The politics of planning and management did not create 'the spatialities of belonging' but they enabled their construction, they allowed the local embodied voices to be heard and to express their daily needs and desires in their community spaces. These ethno-spaces later assist local businessmen to commercialise the area by providing symbolic spatial constructions, which serve the tourists more than the local people.

If we conclude this case by looking at the issues we raised earlier, we can see that again *scale* matters. The fact that the case study area is a neighbourhood in which people from the same origin concentrated assists in determining these constructions of space as focusing on belonging and attachment. *Who designs the city* relates in this case not only to the planners but as much to the people living in the area, to the power of consumerism, to people's demands and the economic forces that made the area look like 'being back in Bangladesh' long before the actual programme was born. Later, the economic forces of local businessmen and the developers' interests joined these processes of city design and perhaps reinforced existing ethnic tendencies. Here *the planning process is defined* in its broader meaning to include not only the actual planning but also the management of the place. Obviously in this process of creating an 'ethnotown' there are many groups involved. This case shows how spaces of belonging are developed from bottom-up initiatives that usually engage in basic cultural needs for food and other cultural items but can be developed further to be constructed as places where ethnicity is used commercially.

CASE STUDY   On belonging and commitment in spatial planning in Cape Town, South Africa

This last case study represents a political and social context in which the politics of planning and development were used to smash and erase people's sense of belonging and memory simply to serve the goals of the apartheid government to ensure the white control and dominance in South African cities. This is probably one of the most cynical and explicit examples of using city planning and expansion to serve racist purposes.

The apartheid government passed the Group Area Act in 1950, a law that forced people into racial groups and made it illegal for people of different races to live in the same area. Based on this law one of the areas to be evicted was District Six in Cape Town, so named in 1867. Originally established as a vibrant mixed community of freed slaves, merchants, artisans, labourers and immigrants, because of its location close to the city and the port it became a centre linking the two together. Its location and the policies of segregation and discrimination pushed the government to evict some 60,000 people between 1966 and 1980. Some of them were evicted to an area known as the Cape Flats, which are actually the townships surrounding Cape Town (see Figure 10.7). The area was bulldozed except for a few churches and mosques. The government wanted

Figure 10.7 Townships in Cape Town, South Africa

to give the area a new image and thus renamed it Zonnebloem, and cynically named the new townships, their streets and blocks after the old street names in District Six. Ironically, the eviction of District Six aroused so much criticism that most of Zonnebloem remains empty today as private entrepreneurs and firms show reluctance to invest in such a controversial area.

In 1994, when the apartheid regime fell the District Six Museum was established 'In order to work with the memories of these experiences' (the District Six Museum of Cape Town). The Museum tells the history of apartheid and its effects on people through intimate stories and memories of people who lived in the area. Memory and documentation are the major source of commemoration and perpetuation of this sense of belonging to the area. The Museum has a map of the district on which each of its former residents can make comments regarding their daily life in the area. It is an explicit and living memory and a clear expression of how the right to belong becomes a crucial part of human and civil rights and in maintaining a civil society.

No doubt, this is an extreme case of a cruel use of the politics of city planning to serve the racist South African policies, but perhaps from the 'extreme' we can learn about the 'regular', those policies of discrimination that in many cases become daily norms in cities and cause abuse of civil rights and the rights to quality of life. In South Africa the replacement of 'old' for 'new' was an excuse for promoting ethnic purification of the city; in other cities it engages economic interests serving both developers and municipalities and leaving the people outside the 'planning and politics' game.

Summary and conclusions

Towards a new understanding of planning for people

This chapter serves as the last step in the process that has been developed in this book of formulating a framework of local knowledge, a framework based on a different understanding of everyday life in the city. This different understanding engages the deconstruction of CBC and their relevance to the way cityscapes are shaped.

Planners and architects discussing these connections raised several themes, which help to identify the contexts in which CBC can be incorporated as local knowledge in the planning process. These points have been largely elaborated in the text. To avoid repetition, I mention them here again but only briefly:

- the credibility and viability of the plan and the role and politics of the planner;
- the scale of the plan;
- the definition and scope of the planning process;
- the mechanisms of civil society and the role of NGOs in promoting civil society norms;
- the 'actors' involved in designing our cities;
- the formal mechanisms – 'community impact assessment'.

The analysis of the three case studies using these themes helps clarify under what circumstances CBC become relevant or irrelevant to planning and development in different political, cultural and social environments.

The three case studies reflect planning events in which citizens as well as NGOs were involved. It seems that citizens' voices get heard when political agendas are targeted towards the realisation of their identities. The case of Cape Town shows how the politics of planning can be used to promote needs other than those of specific groups. The politics of planning in Jerusalem show the same towards the Palestinian living in the city. Citizens' voices get heard when planners' politics are targeted towards people's own needs, and when the plan's viability counts more than its credibility. Their voices get heard when there are mechanisms of civil society to help this process.

The 'old and new' approach in designing urban spaces has been analysed as one of the practical means to maintain a sense of belonging and attachment in the city. This approach is usually implemented with regard to public buildings; but I suggest using this approach more widely with regard to buildings and sites that have local significance for those who live in such places.

Finally, Sandercock (2000) suggests a new way of planning, 'therapeutic planning', which involves negotiation and mediation, facilitation and consensus building. She says: 'this approach is the best model in cases where prior histories of conflict have made more traditional negotiation techniques irrelevant' (p. 28). It seems that at least for Israel and South Africa this model of 'therapeutic planning' can become one of the means to achieve mutual understanding as to the needs and desires of CBC of the parties involved. Therapeutic planning together with the community impact assessment become the vehicle in promoting CBC in planning. They serve as tools for planners to identify the meanings of CBC or other aspects of the quality of life, as the communities themselves perceive them. They help communities to understand the intricacies between what makes them feel comfortable, what increase their sense of belonging and commitment and how urban planning and management can become a vehicle to realise these needs.

CHAPTER 11

Conclusion

This book has sought to create a different body of knowledge regarding the ways and means urban spaces are perceived, used, planned and managed. A specific emphasis has been on formulating the distinction between two sets of knowledge that are involved in the practicalities of everyday life in cities: the local embodied knowledge and the professional planning knowledge.

Local embodied knowledge has been defined as consisting of the ways people interpret and the meanings given to the concepts of comfort, belonging and commitment to the environment as part of their quality of life. This knowledge is 'local' as it represents the daily life of people living in various locales, from the more intimate, private spaces to the more public spaces that people use in their everyday activities. It is a knowledge that is 'embodied' through their identities and physical experiences.

Professional planning knowledge has been defined in this book using Ma Reha's terms (1998) as 'academically generated ideas' that were produced in Western universities and schools of planning from the mid-nineteenth century by various disciplines mainly in the United Kingdom and the United States. These ideas reflected scientific approach to planning expressing the era of modernity and technology. This type of knowledge has been spread all over the world as the ultimate authority of scientific knowledge in planning and thus became the knowledge adequate and appropriate to all places in the world. Its strong expressions can be seen in many cities in Europe and the USA but also in many Third World cities. As largely elaborated in Chapter 2, this process took place primarily by the academic education of students of planning and architecture from Europe, the USA and Third World countries. These students studied in European Universities and internalised the principles of modernist planning into their professional knowledge. This professional knowledge has been in most cases hostile to local traditions, cultural norms and values and perceived them as old fashioned, primitive and sometimes irrelevant. The doubts around the hegemonic nature of the professional planning knowledge in shaping cityscapes around the world started in the early 1960s and motivated the development of alternative approaches to planning that emphasise the incorporation of both professional and local embodied knowledge in the planning process.

I then wanted to examine the links that can be established between these everyday experiences that I term here: 'local embodied knowledge' and the planning practice led by the professional planning knowledge. Incorporating these intimate experiences in the planning process is what moves the planning practice forward and makes the whole process more grounded and therefore more suitable to people's needs and, especially, their desires. I perceived this examination as the end result and as one of the major challenges of this book: How to articulate the local knowledge that has been exposed and analysed in the various chapters into planning practice. We will return to this point in the final part of this chapter.

The distinction between local knowledge and professional planning knowledge reflects power relations, the superiority of white, Western knowledge makes one type of knowledge 'professional', 'scientific', 'universal' and the other type merely 'local' and less significant (Harding 1996). These power relations create *politics of planning and development*, which have different motivations. In London it is the power of capital, which has its dominance in shaping and reshaping cityscapes and has negative effects on the quality of life of communities such as the Bangladeshi in East London. In Jerusalem it is the power of the national ideology, which affects the shaping of politics of planning and development in the city, and has its effects on the quality of life of the Palestinians residents of East Jerusalem.

Quality of life in the city has been another element focused on in the book. The 'city-as-home' approach focuses on the elements identified as quality of life at home with the aim of transforming them into city planning and management: a sense of *comfort*, a sense of *belonging* and a sense of *commitment* to the environment has been narrated by residents of the two cities with regard to six categories of environment: home, building, street, neighbourhood, city centre and the city, public transport and urban parks.

A sense of *comfort* engages physical, social and emotional elements. Its articulation is connected to power relations. Even at 'home', the most private space, a sense of comfort and discomfort results from power relations within family members, especially between men and women. A sense of comfort facilitates determining the boundaries between the 'private' and the 'public' and other symbolic spaces in between these two distinctions. A sense of comfort serves as another boundary marker, this time between 'me' and 'you' or 'us' and 'them'. This means that a sense of comfort is associated with cultural and ethnic homogeneity, especially in neighbourhoods in both cities. A sense of comfort is associated with identity issues as elaborated on below: gender, class and nationality. A sense of comfort in the city has been connected to functions such as cafés and their effects on city life. The bottom line of this analysis has been the articulation of 'the right to comfort in the city', which is linked with many aspects of human and citizen rights because of its connection with power relations and hegemony. As such, the right to comfort has its spatial–physical dimension with regard to the politics of planning and development in the city.

A sense of *belonging* has a multiplicity of meanings. 'The everyday meanings of belonging' do not necessarily emerge out of ideology, religion, nationality or other forms of collective structures but rather belonging is connected to personal, private and intimate daily practices. The different formations of belonging are linked to the different daily practices and identities. Everyday belonging is formed out of repetitive walking practices, which create our 'private city'. Another association of belonging is with the notion of memory, both personal memories of home and childhood and collective imagined memories that comprise the basic formations of nationality and collective identity. Another association of belonging, especially at the neighbourhood level, is with its social, ethnic and religious homogeneity. A sense of belonging is also linked to power relations and control from the very intimate space of home up to the city. Finally, 'the right to belong' has been identified as the right of people of different identities to be recognised and to take part in civil society. The right to belong relates to a situation where one's own rights to equality and the right to preserve identity differences are kept.

A sense of *commitment* is multi-layered too. Commitment can be personal or public, passive or active, physically oriented or socially and ideologically oriented. This made it the most complicated to define of the three components. Commitment has been associated with 'give and take' relations. Commitment to the 'self' is the basis of commitment to other categories of space. The extent to which people feel that the home, the street or the city are committed to them influences the level of their commitment to those environments. A sense of commitment means making choices to live in a city. Purchasing a flat has been perceived as an act of commitment to a city. Commitment is connected to public actions and activities that people are willing to undertake in order to promote or protest against planning and development issues that affect their life. But commitment is also associated with collective memory and with national identity. 'The right to commitment' has been identified as expressing citizen–authority mutual responsibility, or 'give and take' relations. This means the right of people in the city to receive what they are entitled to receive (services and infrastructure) from the authorities and to give what they ought to give to the authorities (tax payments, etc.)

It is possible now to use this in-depth analysis of these three elements as a means of challenging the three questions presented in the introduction.

Are all cities alike?

We started this book with the question of whether daily practices in cities have their similarities in spite of the different histories, economies, social and cultural assets, politics, and different means of imagining their futures. Following this we posed in Chapter 1 a more specific question whether people of similar socio-economic status experience the same everyday life in what are perceived as contrasting cities: the global, world city of London and the holy city of Jerusalem. Are these experiences locally embedded or are there some shared universal experiences of living in urban spaces, whether 'global' or 'holy'? Let us deal with these questions.

Some of the issues highlighted in people's narratives can be seen as similar. *The dominance of power relations* in dictating people's senses of comfort, belonging and commitment has been identified as a common experience for people living in the two cities. Mechanisms of power relations are expressed at all levels of environment, from the home to the city, and in each they have significant effects on the extent to which people feel comfortable, as if they belong, and committed. However, it seems that power relations is a more explicit mechanism in Jerusalem then in London. From people's narratives it seems that urban spaces in Jerusalem are much more controlled than those in London.

The articulation of the 'private' and the 'public' and the distinction between 'us' and 'them', using a sense of comfort as a means of clarification of the various meanings of each category, has been similar as well in both cities. The example of the *building* as a rather complicated environment in terms of the distinctions between what is 'private' and what is 'public' has been identified in people's narratives in both cities. Also the meanings of comfort and belonging as an expression of *social, cultural and ethnic homogeneity* have been similar in the narratives of Londoners and Jerusalemites. It seems that in general senses of comfort and belonging are associated with the desire to live with people similar to oneself. In London this tendency finds its expression in the formation of 'ethno-towns' such as Banglatown, and the narratives of the Bangladeshi people express their sense of comfort in and belonging to the area they live in. In Jerusalem, this tendency is expressed in people's desires to live in segregated neighbourhoods on a national and religious affiliation basis. Despite that, one distinction must be made: that comfort and belonging as associated with homogeneity is more extreme and explicit in Jerusalem than in London.

The discussions on comfort highlighted another shared experience in both cities: *the importance of the café* as an urban institute, which increases a sense of comfort in the city, especially in the city centres but also in residential areas.

Residents of both cities have identified the dominance of memory as a significant component of a sense of belonging. People living in London and Jerusalem have also made the distinction between personal memory and collective memory as creating a sense of belonging. A sense of belonging as a formation of *citizenship* rights had its strong echo in people's narratives in Jerusalem, especially the Palestinian residents of the city. It seems that this aspect of citizenship in the city is much more problematic in Jerusalem than in

London and therefore it had its stronger expressions in people's narratives there.

The dominance of commitment as associated with *public action* is another component similar in people's narratives in London and Jerusalem. People in both cities perceive their public actions as engaged with objections and protests against developers and municipality's intentions to reshape their environment, which might contradict what they perceive as a good quality of life in their locations. In both cities developers' intentions have their negative effects, in particular on low-income neighbourhoods, which again highlight the notion of power relations in shaping cities.

Certain aspects of daily practice are similar in both cities and probably in most cities in the world – the differences relate to the degree and volume of each element; for example, both the Palestinians and Bangladeshi are negatively affected by the power of the hegemony, but these effects are stronger and more explicit in Jerusalem than in London.

The power to influence and to be involved in decision-making processes regarding changes in urban spaces is what makes people feel more comfortable, and increases their sense of belonging and their sense of commitment. These seem to be universally shared experiences, whether one lives in a 'global' or a 'holy' city. This power is also connected to the extent to which people are willing to act publicly and the extent to which their citizenship rights include such access to power and decision making. It seems that the more people are involved in these procedures the more their quality of life is increased, not only because of improvements in physical elements but also and perhaps mostly because they feel involved and they feel that they share the power to shape their environment. In this respect we can conclude that what really affects people's daily practices is their power to be involved in and to have access to resources that the city ought to provide for them.

Do identity issues have any effect on people's everyday experiences?

The analysis of comfort, belonging and commitment discovered that identities certainly matter in people's daily practices. In Chapter 3 we largely elaborated on the existing literature, which highlights how different identities have their effects on people's daily practices in the city, especially gender, age, disability and sexual preferences. In the research two significant identities were focused on: gender and national identity and the book (especially Chapters 5–8) highlighted how these identities shape people's ways of interpreting and the meanings given to comfort, belonging and commitment. *Class* is another identity issue that affects people's sense of comfort. People in both cities prefer to reside with people of the same class, a situation that is created by itself as a result of economic ability. In London, class issue has been more dominant, especially in identifying the boundaries of 'the private' in the streets that lead to people's houses. In several suburban areas in London the streets have been identified as 'semi-private' by the residents of these areas in association with a sense of comfort. We discussed in the previous section the notion of *social, cultural* and *ethnic homogeneity* as increasing a sense of comfort and belonging. This desire for homogeneity, which has been expressed in both cities, can be associated with notions of *exclusion* and *segregation*, which sadly become common experiences in urban areas in the world.

Gendered identity has been identified as very significant in understanding the meanings of comfort, belonging and commitment. The articulation of these three elements is very much gendered and is associated with patriarchal power relations in all categories of the environment. Household gendered divisions of roles were one of the components affecting women's and men's articulations of comfort, belonging and commitment, especially at the home level. Motherhood has been identified as a significant stage in the life cycle, one which contributed to women's increasing sense of comfort and especially belonging to their close environment, in particular the neighbourhood. The role of women's clothing has been discussed in connection with the politics of city management. Here there is a major difference between London and Jerusalem. While in London women's different styles of clothing have become part of daily practices everywhere in the city, in Jerusalem there are public spaces, especially in ultra-orthodox neighbourhoods, that exclude women because of their 'inappropriate' clothing. The difference between the two cities relates to the ideologies, which are the basis of the two cities' governance. They seem to be more 'global' in London in the sense that the city's politics do not allow such exclusionary forms to exist as they do in Jerusalem, where they are more 'local' or 'religious oriented'.

National identity has been another dominant identity in understanding the daily experiences of people in the two cities. National identity is associated with the three elements and more so with 'the right to comfort', 'the right to belong' and 'the right to commitment'. These three definitions are articulated in the definition of 'citizenship in the global city', which respects both citizens' rights to equality and their different ethnic, cultural and national rights. Those rights are not always respected, and the narratives of the Palestinian residents of Jerusalem emphasise this. National identity is very significant in identifying citizen rights abuse in Jerusalem, which in many ways is more obvious than in London.

Putting identity issues at the forefront of the discussion exposes the different daily experiences in the two cities. National identity is more significant in Jerusalem in identifying notions of power and access to resources than in London. Clearly the minorities in the two cities, either ethnic or national, have less power and less access to resources. But it seems that this struggle for equality and difference has been more successful in London than in Jerusalem.

Is local knowledge embodied in people's daily practices valid and viable in the planning practice?

Parts III and IV challenged this question. Part III discussed the use of cognitive maps as a methodology to deepen the understanding of the meanings of comfort, belonging and commitment and Part IV demonstrated the practicalities of incorporating local embodied knowledge into the professional knowledge of planning and provided alternative thinking that moved the discussion forward.

The strong connections between local knowledge of comfort, belonging and commitment and planning practices have been analysed in both Parts III and IV. In Part III the cognitive expressions of CBC have been identified and their relevance to planning practice formulated. Let us summarise the main points that were discussed in the conclusion. First, we discussed the relevance of the symbolic and the physical images that appeared in cognitive temporal maps as indicating sites of memory and belonging, especially the collective ones. Second, we made a distinction in CT maps between images of the 'private' and images of the 'public' and the boundaries between them. These distinctions help planners to understand how residents of a neighbourhood, for example, construct the spaces within their areas of living, that is, which areas they identify as private and which areas as public, and how planners can use this local knowledge to identify the extent to which the boundaries between the two are significant in the process of reshaping these spaces in ways that perhaps better meet the needs of the community members. Third, we identified the contribution of these distinctions to planning practice. The centrality and relationships between the home and the environment can help planners distinguish problems of accessibility and transport in certain parts of the city. Some of the mental maps also indicated the desired relationship between one's own home and the city as increasing a sense of comfort. And fourth, we identified the symbols representing individual identity and collective identity, which again serve as indicators of belonging and attachment in urban spaces. This local knowledge could serve planners whose task is to design or redesign certain parts of the city. Sites representing collective identities can become sites of preservation. As mentioned in Part III, all the above understandings are *scale* oriented. The more detailed the scale of the plan the more meaningful and useful it is to include such local knowledge in the planning process. Also, the practicalities of incorporating local knowledge embedded in CT mapping depend on the *power relationships* between the community, the planner and the authorities.

Finally, let us mention the points of reference that were discussed in Part IV dealing with the connections of CBC and planning practice. These points included the discussion on the *credibility and viability of the plan* and *the role and politics of the planner* in incorporating CBC in planning practice. Clearly there may be a clash between the planner's desire to prepare a viable plan that meets the needs and aspirations of the community but also pay attention to its credibility and its approval by the authorities. The narratives of the planners in Chapter 10 emphasise these dialectics. This point brings us back to the discussion in Chapter 2 on the role of the planner in the planning process, whether the planners perceives themselves as community transformer or professional designer. This view affects the extent to which planners work on the viability or credibility of the plan. The second dimension important to the incorporation of CBC is *the scale of the plan.* As already mentioned, the more detailed the scale the more practical it is to incorporate a planning method that identifies and analyses these components. *The definition and scope of the planning process* is another important point in incorporating CBC. The broader the scope of the planning process beyond the plan making itself the higher the chances of incorporating CBC in both planning and managing of cities. This relates to the next point, which concerns *the scope and mechanisms of civil society and the role of NGOs in promoting civil society norms* – the louder the voices of communities and individuals the greater the chances of promoting elements of quality of life in the city's planning and management. The extent to which CBC are becoming part of planning and management of our cities depends on the *'actors' involved in these processes* and the extent to which governments and municipalities can enforce the incorporation of the quality of life in such processes. Finally, some formal mechanisms such as the *'Community Impact Assessment'* have been formulated as a practical tool for incorporating CBC in planning and managing our cities. Independent planners and community mobilisers who will work with community members themselves on evaluating the impacts of the project on people's own definitions of their quality of life carry out this assessment. Such an assessment could become part of the requirements for obtaining planning and building permits.

This long and detailed journey, which started with challenging existing planning traditions and criticising current formations and dominance of professional knowledge in planning, has come to an end. But I hope this end will be the beginning of other lines of thinking, of formulating new ways of knowledge, insights and different understandings of planning, knowledge and diversity that the book has triggered in its readers.

Bibliography

Abu Lughod, L. (1986) *Veiled Sentiments*, Berkeley, Ca: University of California Press

Abu Lughod, L. (1993) *Writing Women's World*, Berkeley, Ca: University of California Press

Abu Odeh, L. (1993) 'Post-colonial feminism and the veil: thinking the difference' *Feminist Review*, 43, pp. 26–37

Afshar, H. (1996) 'Women and the politics of fundamentalism in iran' in: Afshar, H. (ed.) *Women and Politics in the Third World*, pp. 121–41

Aitken, C.S., Cutter, I.C., Foote, K.E. and Sell, J.L. (1989) 'Environmental perception and behavioral geography' in: Gaile, G.L. and Wilmott, J. (eds) *Geography in America*, Columbus, Oh: Merrill Publishing Company, pp. 218–38

Amnesty International (1999) *Israel and the Occupied Territories: Demolition and dispossession, the destruction of Palestinian homes*, London: Amnesty International

Anderson, B. (1983) *Imagined Communities*, London: Verso

Anderson, K. (1991) *Vancouver's Chinatown*: Racial discourses in Canada, 1875–1980. Montreal, Que: McGill-Queen's University Press

Anteby, L. (1999) 'There's blood in the house: negotiating female rituals of purity among Ethiopian Jews in Israel in: Wasserfull, R. (ed.) *Women and Water: Menstruation in Jewish Life and Law*, Hanover, NH: Brandeis University Press, pp. 168–86

Askenazi, M. and Weingrod, A. (1987) (eds) *Ethiopian Jews and Israel*, New Brunswick, NJ: Transactions Books

Batler, R. (1998) 'Rehabilitating the images of disabled youths' in: Skelton, T. and Valentine, G. (eds) *Cool Places*, London: Routledge, pp. 83–100

Beckett, J. (1996) 'Against nostalgia: place and memory in Myles Lalor's "Oral History"', *Oceania*, 66, pp. 312–27

Bell, V. (1999) 'Performativity and belonging: an introduction' *Theory, Culture and Society*, 16 (2), pp. 1–10

Beller-Hann, I. (1995) 'Women and fundamentalism in Northeast Turkey' *Women: A Cultural Review*, 6, pp. 35–45

Berkowitz, S. (2000) *The Battles over the Holy Sites*, Jerusalem Institute of Research, Jerusalem, Israel (Hebrew)

Bishop, P. (1992) 'Rhetoric, memory, and power: depth psychology and postmodern geography' *Society and Space*, 10 (1), pp. 5–22

Bishop, P. (1995) *An Archetypal Constable*, London: Athlone

Bollens, S. (2000) *On Narrow Ground: Urban policy and ethnic conflict in Jerusalem and Belfast*, New York: State University of New York Press

Borja, J. and Castells, M. (1997) *Local and Global: Management of Cities in the Information Age*, Earthscan, London

Bourdieu, P. (1984) *Distinction: A social critique of the judgement of taste*, London: Routledge

B'tselem (1997) *A Policy of Discrimination: Land Expropriation, Planning and Building in East Jerusalem*, Jerusalem: B'tselem

Bunch, C. (1995) 'Transforming human rights from a feminist perspective' in: Peters, J. and Wolper, A. (eds) *Women's Rights Human Rights*, New York: Routledge, pp. 11–17

Carr, S., Francis, M., Rivlin, L.G. and Stone, A. (1992) *Public Space*, Cambridge University Press

Carter, H. (1981) *The Study of Urban Geography*, London: Edward Arnold

Casey, E.S. (1993) *Getting back into Place*, Bloomington, Ind: Indiana University Press

Castells, M. (1976) *The Urban Question*, London: Edward Arnold

Castells, M. (1997) *The Power of Identity*, London: Blackwell

Central Bureau of Statistics (2000) *Statistical Yearbook of Israel*, Jerusalem (Hebrew)

Cheshin, A., Hutman, B. and Melamed, A. (2000) *Separated and Unequal: The inside story of Israeli rule in East Jerusalem*, Cambridge, Mass: Harvard University Press

Churchman, A. (1993) 'The compatibility of spatial planning to different groups within the population in Israel', Israel 2020 – *Master Plan for Israel for 2000*, Technion, Haifa: The Centre for Urban and Regional Research (in Hebrew)

Churchman, A., Alterman, R., Atzmon, Y., Davidovitch-Merton, R. and Fenster, T. (1996) *Shadow Report for Habitat International – Istanbul*, Jerusalem: Israel's Women's Network

Clark, D. (1996) *Urban World/Global City*, London: Routledge

Clark, J. and Stewart, J. (1997) *The Managerial State*, London: Sage

Crang, M. (1998) *Cultural Geography*, London: Routledge

Colomina, B. (ed.) (1992) *Sexuality and Space*, Princeton, NJ: Princeton Architectural Press

Cosgrove, D. and Jackson, P. (1987) 'New directions in cultural geography', *Area*, 19, pp. 9–11

Cuttis, B. (2001) 'That place where: some thoughts on memory and the city' in: Borden, I., Kerr, J., Rendell, J. and Pivaro, A. (eds) *The Unknown City*, Cambridge, Mass: MIT Press, pp. 54–67

Davidoff, P. (1965) 'Advocacy and pluralism in planning', *Journal of the American Institute of Planning*, 31, pp. 331–38

Davis, M. (1990) *City of Quartz*, London: Verso

de Certeau, M. (1984) *The Practice of Everyday Life*, Berkeley, Ca: University of California Press

Doleve-Gandelman, T. (1990) 'Ethiopia as a lost imaginary: the role of Ethiopian Jewish women in producing the ethnic identity for their immigrant group in Israel' in: Maccannell, F. (ed.) *The Other Perspective of Gender and Culture*, New York: Columbia University Press, pp. 242–57

Downs, R.M. and Stea, D. (1973) 'Cognitive maps and spatial behaviors: processes and products' in: Downs, R.M. and Stea, D. (eds) *Image and Environment: Cognitive mapping and spatial behavior*, Chicago, Ill: Aldine, pp. 8–26

Droogleever, J. (forthcoming) 'Exclusion, coexistence, tolerance and beyond: the elderly in the Netherlands', *Hagar – International Social Science Review*

Duncan, N. (1996) 'Renegotiating gender and sexuality in public and private spaces' in: Duncan, N. (ed.) *BodySpace*, London: Routledge, pp. 127–45

Dwyer, C. (1998) 'Contested identities: challenging dominant representations of young British Muslim Women' in: Skelton, T. and Valentine, G. (eds) *Cool Places*, London: Routledge, pp. 50–65

Eade, J. (ed.) (1997) *Living the Global City: Globalisation as local process*, Routledge, London

Eisenstein, Z. (1996) 'Women's publics and the search for new democracies', paper presented at the conference: Women, Citizenship and Difference, London, July 1996

Fainstein, S. (1994) *The City Builders*, Oxford: Blackwell

Fainstein, S. (1996) 'Planning in a different voice' in: Campbell, S. and Fainstein, S. (eds) *Readings in Planning Theory*, Oxford: Blackwell, pp. 456–460

Fainstein, S. (2000) 'New directions in planning theory', *Urban Affairs Review*, 35 (4), pp. 451–78

Faludi, A. (1997) 'A planning doctrine for Jerusalem?', *International Planning Studies*, 2 (1), pp. 83–103

Fenster, T. (1996) 'Ethnicity and citizen identity in development and planning for minority groups', *Political Geography*, 15 (5), 1996, pp. 405–18

Fenster, T. (1997) 'Relativism vs universalism in planning for minority women in Israel', *Israel Social Science Research*, 2 (2), pp. 75–96

Fenster, T. (1998) 'Ethnicity, citizenship and gender: the case of Ethiopian immigrant women in Israel', *Gender, Place and Culture*, 5 (2), pp. 177–89

Fenster, T. (1999a) 'Space for gender: cultural roles of the forbidden and the permitted', *Environment and Planning D: Society and Space*, 17, pp. 227–46

Fenster, T. (1999b) 'Culture, human rights and planning (as control) for minority women in Israel' in: Fenster, T. (ed.) *Gender, Planning and Human Rights*, London: Routledge, pp. 39–54

Fenster, T. (1999c) 'On particularism and universalism in modernist planning: mapping the boundaries of social change', *Plurimondi*, 2, pp. 147–68

Fenster, T. (2001) 'Planning, culture, knowledge and control: minority women in Israel' in: Yiftachel, O., Little, J., Hedgcock, D. and Alexander, I. (eds) *The Power Of Planning*, Dordrecht: Kluwer, pp. 77–89

Fenster, T. (2002) 'Planning as control: cultural and gendered manipulation and misuse of knowledge', *Hagar – International Social Science Review* (1), pp. 67–84

Fenster, T. (forthcoming) 'Nostalgia, cognitive temporal maps and spatial planning'

Forester, J. (1989) *Planning in the Face of Power*, Berkeley, Ca: University of California Press

Forester, J. (1999) *The Deliberative Practitioner*, Cambridge, Mass: MIT Press

Forman, C. (1989) *Spitalfields: A Battle for Land*, London: Hilary Shipman

Forty, A. (2000) *Words and Buildings: A vocabulary of modern architecture*, London: Thames & Hudson

Foucault, M. (1980) *Power/Knowledge*, New York: Harvester Wheatsheaf

Friedmann, J. (1973) *Retracking America*, New York: Doubleday Anchor

Fullilove, M.T. (1996) 'Psychiatric implications of displacement: contributions from the psychology of place', *The American Journal of Psychiatry*, 153 (12), pp. 1516–22

Giddens, A. (1979) *Central Problems in Social Theory*, London: Macmillian

Giddens, A. (1981) *The Contemporary Critique of Historical Materialism: Power, property and the state*, London: Macmillan

Giddens, A. (1994) 'Living in a post-traditional society' in: Beck, U., Giddens, A. and Lash, S. (eds) *Reflective Modernisation*, Cambridge: Polity Press, pp. 56–109

Gleeson, B. (1998) 'Justice and the disabling city' in: Fincher, R. and Jacobs, J. (eds) *Cities of Difference*, New York: Guildford Press, pp. 89–119

Gobster, P.H. (1998) 'Urban parks as green walls or green magnets: interracial relations in neighbourhood boundary parks', *Landscape and Urban Planning*, 41, pp. 43–55

Golledge, R.G. and Stimson, R.J. (1997) *Spatial Behavior*, New York: Guildford Press

Gregory, D. and Urry, J. (1985) *Social Relations and Spatial Structures*, London: Macmillian

Hall, S. (1996) 'Introduction: Who Needs Identity?' in Hall, S. and du Gay, P. (eds) *Questions of Cultural Identity*, Thousand Oaks: Sage, pp. 1–17

Halseth, G. and Doddridge, J. (2000) 'Children's cognitive mapping: a potential tool for neighbourhood planning', *Environment and Planning B: Planning and Design*, 27, pp. 565–82

Harding, S. (1996) 'Gendered ways of knowing and the "epistemological crisis" of the West' in: Goldberg, N., Tarule, J., Clinchy, B. and Belenky, H. (eds) *Knowledge, Difference, and Power*, New York: Basic Books, pp. 431–54

Harley, J. (1988) 'Maps, knowledge, and power' in: Cosgrove, D. and Daniels, S. (eds) *The Iconography of Landscape*, Cambridge University Press, pp. 277–312

Harrell, W.A., Bowlby, J.W. and Hall-Hoffarth, D. (2000) 'Directing wayfinders with maps: the effects of gender, age, route complexity, and familiarity with the environment', *The Journal of Social Psychology*, 140 (2), pp. 169–78

Harvey, D. (1973) *Social Justice and the City*, London: Edward Arnold

Harvey, D. (1985) *The Urbanization of Capital*, Oxford: Blackwell

Harvey, D. (1987) *Urban Land Economies*, London: Macmillian

Harvey, D. (1989) *The Conditions of Modernity*, Oxford: Blackwell

Harvey, D. (1996) *Justice, Nature and the Geography of Difference*, Oxford: Blackwell

Harvey, D. (1999) 'Frontiers of insurgent planning', *Plurimondi*, 2, pp. 269–86

Hasson, S., Shcory, N. and Adiv, H. (1995) *Neighbourhood Governance: The concept and the method*, Jerusalem: The Jerusalem Institute for Israeli Studies, Research Series No. 68

Healey, P. (1992) 'A planner's day: knowledge and action in communicative perspective', *Journal of the American Planning Association*, 58 (1), pp. 9–20

Healey, P. (1997) *Collaborative Planning*, London: Macmillan

Hillier, J. (1998) 'Representation, identity, and the communicative shaping of place' in: Light, A. and Smith, M.J. (eds) *The Production of Public Space*, Oxford: Rowman & Littlefield, pp. 207–32

hooks, b. (1991) *Yearning: Race, gender, and cultural politics*, London: Turnaround

hooks, b. (2001) 'City living: love's meeting place' in: Borden, I., Kerr, J., Rendell, J. and Pivaro, A. (eds) *The Unknown City*, Cambridge: MIT Press, pp. 436–41

Hurtado, A. (1996) 'Strategic suspensions: feminists of color theorize the production of knowledge' in: Goldberg, N., Tarule, J., Clinchy, B. and Belenky, M. (eds) *Knowledge, Difference, and Power*, New York: Basic Books, pp. 372–92

Huxley, M. (1994) 'Planning as a framework of power: utilitarian reform, enlightment logic and the control of urban spaces' in: Ferber, S., Healy, C. and McAuliffe, C. (eds) *Beasts of Suburbia*, Melbourne University Press, pp. 148–69

Imrie, R. (1999) 'The implications of the "New Managerialism" for planning in the millennium' in: Allmendinger, P. and Chapman, M. (eds) *Planning Beyond 2000*, Chichester: John Wiley, pp. 107–22

Innes, J. (1998) 'Planning theory's emerging paradigm: communicative action and inter-active practice', *Journal of Planning Education and Research*, 14 (3), pp. 183–90

Jackson, P. (1989) *Maps of Meanings*, London: Routledge

Jackson, P. (1998) 'Domesticating the street: the contested spaces of the high street and the mall' in: Fyfe, N.R. (ed.) *Images of the Street*, London: Routledge, pp. 176–91

Jacobs, J. (1961) *The Death and Life of the Great American Cities*, New York: Vintage Books

Jacobs, J.M. (1996) *Edge of Empire: Postcolonialism and the city*, London: Routledge

The Jerusalem Institute for Israeli Studies (2001) *Jerusalem: Facts and Figures*

Kallus, R. and Law-Yone, H. (2000) 'National home/personal home: the role of public housing in the shaping of space', *Theory and Criticism*, 16, pp. 153–80

Kamaluddin, K. (2000) *Drug Awareness Education*, Peabody Trust, Tower Hamlets, London

Kanter, R.M. (1995) *World Class: Thriving locally in the global economy*, New York: Simon & Schuster

Katz, C. and Monk, J. (1993) *Full Circles: geographies of women over the life course*, London: Routledge

Kettle, J. and Moran, C. (1999) 'Social housing and exclusion' in: Allmendinger, P. and Chapman, M. (eds) *Planning Beyond 2000*, Chichester: John Wiley

Kilian, T. (1998) 'Public and private, power and space' in: Light, A. and Smith, J. (eds) *Philosophy and Geography II: The production of public space*, London: Rowman & Littlefield, pp. 115–34

Klein, M. (2001) *Jerusalem: A contested city*, London: Hurst & Company/Jerusalem: The Jerusalem Institute for Israeli Studies (Hebrew)

Kofman, E. (1995) 'Citizenship for some but not for others: spaces of citizenship in con-temporary Europe', *Political Geography*, 14, pp. 121–37

Lane, M. (1999) 'Indigenous people and resource planning in Northern Australia: re-shaping or reproducing existing practice?', *Plurimondi*, 2, pp. 181–92

Leach, E. (1976) *Culture and Communication*, Cambridge University Press

Leach, N. (2002) 'Belonging: towards a theory of identification with space' in: Hillier, J. and Rooksby, E. (eds) *Habitus: A sense of place*, Aldershot: Ashgate, pp. 281–98

Lee, R.G. (1972) 'The social definition of outdoor recreation places' in: Bruch, W. (ed.) *Social Behavior, Natural Resources and the Environment*, New York: Harper & Row, pp. 68–84

Lee, T. (1968) 'Urban neighborhood as a socio-spatial schemata', *Human Relations*, 21, pp. 241–68

Lefebvre, H. (1992) *The Production of Space*, Oxford: Blackwell

Lewando-Hundt, G. (1984) 'The exercise of power by Bedouin women in the Negev' in: Marx, E. and Shmueli, A. (eds) *The Changing Bedouin*, New Brunswick, NJ: Transactions Books, pp. 83–123

Lieblich, A., Tuval-Mashiach, R. and Zilber, T. (1998) *Narrative Research: Reading, analysis and interpretation*, London: Sage

Little, J. (1999) 'Women, planning and local central relations in the UK' in: Fenster, T. (ed.) *Gender, Planning and Human Rights*, London: Routledge, pp. 25–38

London Plan, May, 2001: www.london.gov.uk/mayor/case_for_london/index/htm

Lynch, K. (1960) *The Image of the City*, Cambridge, Mass: MIT Press

Ma Reha, Z. (1998) 'White knowledge and Western superiority', paper Presented at the Annual Conference of the Institute of British Geographers, January 1998

McDowell, L. (1999) 'City life and difference: negotiating diversity' in: Allen, J., Massey, D. and Pryke, M. (eds) *Unsettling Cities*, London: Routledge, pp. 95–136

Macleod, A. (1990) *Accommodating Protest*, New York: Columbia University Press

Macnaghten, P., Grove-White, J., Jacobs, M. and Wynne, B. (1995) *Public Participation and Sustainability: Indicators, institutions, participation*, Preston: Lancashire County Council

Madanipour, A. (1996) *Design of Urban Space*, Chichester: John Wiley

Madanipour, A. (2001) 'How relevant is "planning by neighbourhoods" today?', *Town Planning Review*, 2, pp. 171–91

Madge, C. (1997) 'Public parks and the geography of fear', *Tijdschrif voor Economische en Sociale Geograpie*, 88 (3), pp. 237–50

Malki, R. (2000) 'The physical planning of Jerusalem' in: Ma'oz, M. and Nusseibeh, S. (eds) *Jerusalem: Points of friction – and beyond*, The Hague: Kluwer, pp. 25–56

Ma'oz, M. (2000) 'The future of Jerusalem: Israeli perceptions' in: Ma'oz, M. and Nusseibeh, S. (eds) *Jerusalem: Points of friction – and beyond*, The Hague: Kluwer, pp. 1–6

Markus, T. and Cameron, D. (2002) *The Words Between the Spaces: Buildings and language*, London: Routledge

Marshall, T.H. (1950) *Citizenship and Social Class*, Cambridge University Press

Marshall, T.H. (1975) *Social Policy in the Twentieth Century*, London: Hutchinson

Marshall, T.H. (1981) *The Right to Welfare and Other Essays*, London: Heinemann

Massey, D. (1994) *Space, Place and Gender*, Cambridge: Polity Press

Massey, D., Allen, J. and Pile, S. (eds) (1999) *City Worlds*, London: Routledge/ Buckingham: The Open University

Mernissi, F. (1975) *Beyond the Veil*, Bloomington, Ind: Indiana University Press

Mernissi, F. (1987) *Beyond the Veil*, Bloomington, Ind: Indiana University Press

Mernissi, F. (1991) *Women and Islam: A historical and theoretical inquiry*, Oxford: Blackwell

Mesch, G.S. and Manor, O. (1998) 'Social ties, environmental perception, and local attachment' *Environment and Behaviour*, 30 (4), 504–19

MFA (Ministry of Foreign Affairs) (2002) 'Victims of Palestinian violence and terrorism since September 2000', www.israel-mfa.gov.il

Ministry of Absorption (1985) *The Absorption of Ethiopian Jews – Master Plan*, Jerusalem

Ministry of Absorption (1991) *A Plan for Absorption of the Ethiopian Jews – Second Wave*, Jerusalem

Ministry of the Interior (1999) *The National Master Plan for Israel (Tama/35)*, Jerusalem (Hebrew)

Mitchell, D. (1995) 'There's no such thing as culture: towards a reconceptualization of the idea of culture in geography', *Transactions of the Institute of British Geographers*, 20, pp. 102–16

Mitchell, D. (2000) *Cultural Geography*, Oxford: Blackwell

Mohanty, C.T. (1991) 'Under Western eyes: feminist scholarship and colonial discourses', in: Mohanty, C.T., Russo, A. and Torres, L. (eds) *Third World Women and the Politics of Feminism*, Bloomington, Ind: Indiana University Press, pp. 51–80

Moore, H. (1996) *Space, Text and Gender*, New York: Guildford Press

Moser, C. (1993) *Gender, Planning and Development*, London: Routledge

Mumford, L. (1945) 'The neighbourhood and the neighbourhood unit', *Town Planning Review*, 24, pp. 256–70

Newsweek (1998) 'Where Whirred is a Way of Life', pp. 38–43

Painter, J. and Philo, C. (1995) 'Spaces of citizenship: an introduction', *Political Geography*, 14, pp. 107–20

Pankhurst, H. (1992) *Gender, Development and Identity*, London: Zed Books

Pateman, C. (1988) *The Sexual Contract*, Cambridge: Polity Press

Pateman, C. (1989) *The Disorder of Women*, Cambridge: Polity Press

Patton, M.Q. (1987) *How to Use Qualitative Methods in Evaluation*, Newbury Park, California: Sage

Pile, S. (1996) *The Body and the City*, London: Routledge

Pitkin, D. (1993) 'Italian urbanscape: intersection of private and public' in: Rotenberg, R. and McDonogh, G. (eds) *The Cultural Meaning of Urban Space*, London: Bergin & Garvey, pp. 95–102

Raban, J. (1998) *Soft City*, London: Harvill Press

Rabinowitz, D. (1995) 'Widening the path to rescue the Mulatto women: an Israeli forum,' *Theory and Critique*, 7, pp. 5–19 (Hebrew)

Ram, U. (1999) 'Between colonialism and consumerism: liberal post-Zionism in the global age', Negev Center For Regional Development, Working Paper no. 12, Beer Sheva: Ben Gurion University (Hebrew)

Rao, A. (1995) 'The politics of gender and culture in international human rights discourse' in: Peters, J. and Wolper, A. (eds) *Women's Rights Human Rights*, New York: Routledge, pp. 167–75

Read, P. (2000) *Belonging: Australians, Place and Aboriginal Ownership*, Cambridge University Press

Rendell, J. (1998) 'Displaying sexuality: gendered identities and the early nineteenth-century street' in: Fyfe, N.R. (ed.) *Images of the Street*, London: Routledge, pp. 75–91

Romann, M. and Weingrod, A. (1991) *Living Together Separately*, Princeton, NJ: Princeton University Press

Rose, G. (1997) 'Situating knowledge: positionality, reflexivities and other tactics', *Progress in Human Geography*, 21 (3), pp. 305–20

Safier, M. (1990) Personal Communication

Salamon, H. (1993) '*Beta Israel and their Christian neighbours in Ethiopia: an analysis of central perceptions at different levels of cultural articulation*', unpublished PhD dissertation, Jerusalem, Hebrew University (Hebrew)

Sandercock, L. (1998a) *Towards Cosmopolis*, London: John Wiley

Sandercock, L. (1998b) 'Introduction: framing insurgent historiographies for planning' in: Sandercock, L. (ed.) *Making the Invisible Visible*, Berkeley, Ca: University of California Press, pp. 1–36

Sandercock, L. (1999) 'Introduction. Translations: from insurgent planning practices to radical planning discourses', *Plurimondi*, 2, pp. 37–46

Sandercock, L. (2000) 'When strangers become neighbours: managing cities of difference', *Planning Theory and Practice*, 1 (1), pp. 13–30

Sandercock, L. (2003) *Cosmopolis 2: Mongrel cities in the 21st century*, London: Continuum

Sandercock, L. and Forsyth, A. (1996) 'Feminist theory and planning theory: the epistemological linkages' in: Campbell, S. and Fainstein, S. (eds) *Readings in Planning Theory*, Oxford: Blackwell, pp. 471–74

Sassen, S. (1991) *Cities in a World Economy*, London and New Delhi: Pine Forge Press

Sassen, S. (1994) *The Global City*, Princeton University Press

Sharkansky, I. (1996) *Governing Jerusalem*, Detroit: Wayne State University Press

Sibley, D. (1995) *Geographies of Exclusion*, London: Routledge

Sibley, D. (1998) 'Problemitizing exclusion: reflections on space, difference and knowl-edge', *International Planning Studies*, 3 (1), pp. 93–100

Smith, N. (1994) 'Marxist Geography' in: Gregory, D., Johnston, R. and Smith, D. (eds) *The Dictionary of Human Geography*, Oxford: Blackwell, pp. 471–74

Soja, A. (1989) *Postmodern Geographies*, London: Verso

Solecki, W.D. and Welch, J.M. (1995) 'Urban parks: green spaces or green walls?', *Landscape and Urban Planning*, 32, pp. 93–106

Spain, D. (1993) *Gendered Spaces*, Chapell Hill and London: The University of North Carolina Press

Spitalfields Community Development Plan, 1989–90: London Boroughs of Hackney and Tower Hamlets, CLAWS Good Practice Guide, London

Strüder, I. (2000) 'Social exclusion and its physical/material representation: the example of elderly women in Germany', paper presented at the International Workshop of the Gender and Geography Commission, International Geographical Union, Tel Aviv University, June 2000

Sullivan, D. (1995) 'The public/private distinction in international human rights law' in: Peters, J. and Wolper, A. (eds) *Women's Rights Human Rights*, New York: Routledge

The British Museum (2001) *Exploring the City: The Foster Studio* (exhibition pamphlet)

Thornley, A. (ed.) (1992) *The Crisis of London*, London: Routledge

Tohidi, N. (1991) 'Gender and Islamic Fundamentalism: feminist politics in Iran' in: Mohanty, C.T., Russo, A. and Torres, L. (eds) *Third World Women and the Politics of Feminism*, Bloomington, Ind: Indiana University Press, pp. 251–70

Tolman, E. (1948) 'Cognitive maps in rats and men', *Psychological Review*, 56, pp. 144–55

Valentine, G. (1993) 'Negotiating and managing multiple sexual identities: lesbian–time space strategies', *Transactions of the Institute of British Geographers*, 18, pp. 237–48

Valentine, G. (1998) 'Food and the production of the civilized street' in: Fyfe, N.R. (ed.) *Images of the Street*, London: Routledge, pp. 192–204

Valentine, G., Skelton, T. and Chambers, D. (eds) (1998) 'Cool Places: an Introduction to Youth and Youth Cultures' in Skelton, T. and Valentine, G. (eds) *Cool Places*, London: Routledge, pp. 83–100

Webster, C. (2001) 'Gated cities of tomorrow', *Town Planning Review*, 2, pp. 149–69

Westheimer, R. and Kaplan, S. (1992) *Surviving Salvation*, New York University Press

Wilson, E. (1991) *The Sphinx in the City*, Berkeley, Ca: University of California Press

Wilson, E. (1997) 'Looking backward: nostalgia and the city' in: Westwood, S. and Williams, J. (eds) *Imaging Cities: Scripts, signs, memory*, London: Routledge

Yediot Hachronot newspaper (1998) '360 thousand Israelis surf the Internet' (Hebrew)

Yiftachel, O. (1995) 'Planning as control: policy and resistance in a deeply divided soci-ety', *Progress in Planning*, 44 (2), Ch. 2

Yiftachel, O. (2000) 'Ethnocracy and its discontents: minorities, protests and the Israeli polity', *Critical Inquiry*, 26 (4), pp. 725–56

Yiftachel, O. and Yacobi, H. (2002) 'Planning a bi-national capital: should Jerusalem remain united?', *Geoform*, 33, pp. 137–45

Young, I.M. (1990) *Justice and the Politics of Difference*, Princeton University Press

Yuval Davis, N. (1997) *Gender and Nation*, London: Sage

Yuval Davis, N. (2000) 'Multi-layered Citizenship and the Boundaries of the "Nation–State"' in *Hagar – International Social Review*, pp. 112–27

Yuval Davis, N. (2003) 'Belongings: in between the indegene and the diasporic' in: Ozkirimli, U. (ed.) *Nationalism in the 21ˢᵗ Century*, Basingstoke: Macmillian

Zukin, S. (1997) *The Cultures of Cities*, Oxford: Blackwell

Index